Lecture Notes in Computer Science 14373

Founding Editors

Gerhard Goos
Juris Hartmanis

The series Lecture Notes in Computer Science (LNCS), including its subseries Lecture Notes in Artificial Intelligence (LNAI) and Lecture Notes in Bioinformatics (LNBI), has established itself as a medium for the publication of new developments in computer science and information technology research, teaching, and education.

LNCS enjoys close cooperation with the computer science R & D community, the series counts many renowned academics among its volume editors and paper authors, and collaborates with prestigious societies. Its mission is to serve this international community by providing an invaluable service, mainly focused on the publication of conference and workshop proceedings and postproceedings. LNCS commenced publication in 1973.

Seyed-Ahmad Ahmadi · Sérgio Pereira
Editors

Graphs in Biomedical Image Analysis, and Overlapped Cell on Tissue Dataset for Histopathology

5th MICCAI Workshop, GRAIL 2023 and 1st MICCAI Challenge, OCELOT 2023
Held in Conjunction with MICCAI 2023
Vancouver, BC, Canada, September 23, and October 4, 2023
Proceedings

 Springer

Editors
Seyed-Ahmad Ahmadi 🔟
Nvidia
Munich, Germany

Sérgio Pereira
Lunit Inc.
Seoul, Korea (Republic of)

ISSN 0302-9743   ISSN 1611-3349 (electronic)
Lecture Notes in Computer Science
ISBN 978-3-031-55087-4   ISBN 978-3-031-55088-1 (eBook)
https://doi.org/10.1007/978-3-031-55088-1

This Springer imprint is published by the registered company Springer Nature Switzerland AG
The registered company address is: Gewerbestrasse 11, 6330 Cham, Switzerland

Paper in this product is recyclable.

GRAIL 2023 Preface

The Fifth International Workshop on **GR**aphs in biomedic**A**l **I**mage ana**L**ysis (GRAIL 2023) was organized as a satellite event of the 26th International Conference on Medical Image Computing and Computer Assisted Intervention (MICCAI 2023) in Vancouver, Canada. After a hybrid event in Singapore 2022, we were assigned a fully virtual slot in 2023. This allowed us to make registration and attendance fully free for all participants of our event.

Our GRAIL workshop provides a unique opportunity to meet and discuss both theoretical advances in graphical methods as well as the practicality of such methods when applied to complex biomedical imaging problems. Simultaneously, the workshop seeks to be an interface to foster future interdisciplinary research, including signal processing and machine learning on graphs.

Graphs are powerful mathematical structures that provide a flexible and scalable framework to model unstructured information, objects and their interactions in a readily interpretable fashion. Graph theory provides a solid mathematical foundation for models and algorithms, such as spectral analysis, dimensionality reduction and network analysis. Since 2017, geometric deep learning has married the field of graph signal processing with the flexibility and rapid advancements of deep neural architectures. Since then, applications of graph neural networks (GNNs) in medicine have been steadily increasing, ranging from medical imaging and shape understanding, brain connectomics, population models and patient multi-omics to discovery and design of novel drugs and therapeutics.

In the leading computer vision and ML conferences (CVPR, ICLR, NeurIPS), GNNs were among the emerging topics in 2022 and further increased in interest in 2023. With our GRAIL workshop, we aim to provide the MICCAI community a similar platform for understanding and application of graph-based models as versatile and powerful tools in biomedical image analysis and beyond. The methods scope included a wide range of topics, including but not limited to: deep/machine learning on graphs, probabilistic graphical models for biomedical data analysis, signal processing on graphs for biomedical image analysis, explainable AI (XAI) methods in geometric deep learning, big data analysis with graphs, graphs for small data sets, semantic graph research in medicine, modeling and applications of graph symmetry/equivariance and graph generative models. Application areas included: emerging topics such as graph approaches for intra-operative support, graph analysis of brain networks, and graph representations in pathology imaging, graph-based shape modeling and image segmentation/classification; the scope was also extended to graph and transformer methods for large-scale patient population analyses and multimodal/multi-omics fusion.

The GRAIL 2023 proceedings contain nine high-quality papers of at least 10 pages, which were pre-selected through rigorous double-blind peer-review. Submissions were reviewed on average by at least three members of the reviewing board and Program Committee, comprising 20 experts on graphs in biomedical image analysis, each doing two or more reviews. The accepted manuscripts covered a wide set of methods and applications.

As in previous years, one of the primary domains of imaging-related graph methods was cell graphs or patch graphs in digital pathology and Whole Slide Image (WSI). Another recurring topic at GRAIL is population graphs, this year for brain age regression, or concerning graph assessment metrics. Segmentation was covered in the context of vessel structure graphs. Compared to last year, only one accepted work considered brain connectomes. Emerging topics were graph-based generative AI approaches: namely for graph diffusion, synthetic graph-based lesions, or radiological scene or knowledge graphs.

Special congratulations go to Best Paper Award winners at GRAIL 2023: Vishwesh Ramanathan and Anne Martel were recognised for their paper on "Self Supervised Multi-View Graph Representation Learning in Digital Pathology".

The event was complemented with three exciting keynote talks from world-renowned experts: Matthias Fey from Kumo.AI gave a talk on the popular GNN framework PyG, titled "PyTorch Geometric - Practical Realization of Graph Neural Networks". Dorina Thanou from École Polytechnique Fédérale de Lausanne gave a talk on "Informed machine learning for biology and medicine: A graph representation perspective". Finally, Ranjie Liao from University of British Columbia gave a talk on "Graph Neural Networks Meet Transformers: Some Insights in Representation Learning and Generative Modeling". We cordially thank all the keynote speakers for sharing their insights at our event, and inspiring emerging researchers in this domain at MICCAI.

Most importantly, we wish to thank all the GRAIL 2023 authors for their submissions and participation, and the members of the Program Committee, the numerous reviewers and once more the keynote speakers for their valuable contributions and commitment to the workshop.

The proceedings of our workshop are published as a joint LNCS volume alongside the MICCAI 2023 workshop "OCELOT 2023: Cell Detection from Cell-Tissue Interaction", which aligns well with our papers on the topic of graph-based analytics of digital pathology and WSI imaging. For further information on our event, please visit the GRAIL website (http://grail-miccai.github.io/).

December 2023

Seyed-Ahmad Ahmadi
Anees Kazi
Kamilia Mullakaeva
Bartłomiej W. Papież

OCELOT 2023 Preface

OCELOT (Overlapped CELl On Tissue) 2023: Cell Detection from Cell-Tissue Interaction challenge was organized as a satellite event of the 26th International Conference on Medical Image Computing and Computer Assisted Intervention (MICCAI 2023) in Vancouver, Canada. The challenge was open to participants from June 5, 2023, to August 4, 2023, on the Grand Challenge platform.

Cell detection and classification in histology images is one of the most important tasks in computational pathology. To better locate and classify cells, detailed morphological characteristics such as color and shape are crucial. Consequently, cell detection datasets are typically collected at high magnification but small Field-of-View (FoV). This may have the side-effect of the cell detection model overly relying on appearance details, without understanding the broader context. However, the larger context can provide information about how cells are arranged and grouped together to form high-level tissue structures.

In practice, expert annotators (pathologists) zoom out to understand the broader tissue structures and zoom in to better classify individual cells while taking into account the context information. For example, this behavior is observed in the task of quantifying tumor purity, where the ratio of tumor/non-tumor cells is used, and the context is required to understand whether a given area in the Whole Slide Image (WSI) is cancer. The behavior of pathologists can be transferred to deep learning, for instance, by leveraging data for cell detection tasks at high magnification and data for tissue segmentation tasks at lower magnification. This type of approach would allow the model to share knowledge across different tasks and FoVs. However, to train such an approach, a combined dataset with cell-tissue overlapping regions is required; unfortunately, most existing datasets only target a single task, either cell detection or tissue segmentation. This hinders the development of methods that can efficiently tackle and benefit from both tasks simultaneously.

In the OCELOT 2023 challenge, we aimed to provide data and promote research on how to utilize cell-tissue relationships for better cell detection. To do so, we collected a dataset that contains cell and tissue annotations in small (cell) and large (tissue) FoV patches, respectively, with overlapping regions. Unlike typical cell detection tasks/challenges, participants could utilize tissue patches and annotations for the purpose of boosting cell detection performance.

After the challenge had been successfully finished with 14 submissions in the test phase, the top-5 participants were asked to share their insights at the online event[1], held on September 11, 2023. The speakers shared their submissions in detail: how to generate appropriate labels from the dataset, filter data, model cell-tissue relationships, train the cell and tissue model, and post-process predictions. Also, every participant who submitted to the test phase was asked to submit manuscripts for this LNCS volume.

---

[1] https://www.youtube.com/watch?v=1VAK3t3euY4.

The OCELOT 2023 proceedings contain six high-quality papers, each selected via a rigorous single-blind peer-review process involving 2 reviewers and 1 meta-reviewer from the Program Committee, composed of experts on AI in the pathology domain. The accepted manuscripts cover a wide range of methods utilizing tissue information for better cell detection, in the sense of training strategy, model architecture, and especially how to model cell-tissue relationships. The papers propose various methods to model cell-tissue relationships; rule-based fusing of cell prediction with tissue prediction, concatenating tissue prediction to the cell input, and injecting tissue prediction into the cell model. The variety of methods presented in the papers shows the effectiveness of utilizing tissue information for cell detection improvement.

We wish to thank all the OCELOT 2023 challenge participants for their interest, the authors of the proceedings for their manuscript submission and willingness to share their insights, the members of the Program Committee, the reviewers, the speakers at the online event, and the annotators who contributed to creating the OCELOT dataset. Also, we are very grateful to Lunit Inc. for their support and for permitting the release of the OCELOT dataset to the research community.

The proceedings of our challenge are published as a joint LNCS volume alongside the Fifth International Workshop on GRaphs in biomedicAl Image anaLysis (GRAIL 2023) organized in conjunction with MICCAI. In addition to the papers, slides presented during the online event and event recording were made publicly available on the OCELOT 2023 Grand Challenge website (https://ocelot2023.grand-challenge.org/ocelot2023/).

December 2023

Sérgio Pereira
JaeWoong Shin
Jeongun Ryu
Aaron Valero
Mohammad Mostafavi
Seonwook Park

Organization

General Chair of GRAIL 2023

Seyed-Ahmad Ahmadi   NVIDIA, Germany

Program Committee Chairs of GRAIL 2023

Seyed-Ahmad Ahmadi   NVIDIA, Germany
Anees Kazi   Harvard Medical School, USA
Kamilia Mullakaeva   Technical University of Munich, Germany
Bartłomiej W. Papież   University of Oxford, UK

Program Committee of GRAIL 2023

Bintsi Kyriaki-Margarita   Imperial College London, UK
Czempiel Tobias   Technical University of Munich, Germany
Farshad Azade   Technical University of Munich, Germany
Fischer Markus   Deloitte, Germany
Gopinath Karthik   Harvard Medical School, USA
Holm Felix   Technical University of Munich, Germany
Keicher Matthias   Technical University of Munich, Germany
Li Gang   UNC Chapel Hill, USA
Loddo Andrea   University of Cagliari, Italy
Mueller Tamara   Technical University of Munich, Germany
Nebli Ahmed   Forschungszentrum Juelich, Germany
Pellegrini Chantal   Technical University of Munich, Germany
Popov Pavel   Deloitte, Germany
Rekik Islem   Imperial College London, UK
Rieger Franz   Max Planck Institute for Biological Intelligence,
                             Germany
Scheinost Dustin   Yale University, USA
Starck Sophie   Technical University of Munich, Germany
Xiao Rui   Technical University of Munich, Germany
Yap Pew-Thian   UNC Chapel Hill, USA
Yeganeh Yousef   Technical University of Munich, Germany

General Chair of OCELOT 2023

Sérgio Pereira   Lunit, South Korea

Program Committee of OCELOT 2023

Jeongun Ryu (Chair)   Lunit Inc., Republic of Korea
JaeWoong Shin (Chair)   Lunit Inc., Republic of Korea
Biagio Brattoli   Lunit Inc., Germany
Jinhee Lee   Lunit Inc., Republic of Korea
Wonkyung Jung   Lunit Inc., Republic of Korea
Chan-Young Ock   Lunit Inc., Republic of Korea
Donggeun Yoo   Lunit Inc., Republic of Korea
Aaron Valero Puche (Chair)   Lunit Inc., Republic of Korea
Seonwook Park   Lunit Inc., Republic of Korea
Mohammad Mostafavi   Lunit Inc., Republic of Korea
Sérgio Pereira (General Chair)   Lunit Inc., Republic of Korea
Soo Ick Cho   Lunit Inc., Republic of Korea
Kyunghyun Paeng   Lunit Inc., Republic of Korea

Sponsor of OCELOT 2023

We are very grateful to our sponsor, Lunit.Inc, for their invaluable support in organizing the OCELOT 2023 Challenge and awarding the prizes.

Contents

OCELOT 2023

GRAIL 2023

SCOPE: Structural Continuity Preservation for Retinal Vessel Segmentation

Yousef Yeganeh[1,2], Göktuğ Güvercin[1], Rui Xiao[1], Amr Abuzer[1], Ehsan Adeli[3], Azade Farshad[1,2(✉)], and Nassir Navab[1,4]

[1] Technical University of Munich, Munich, Germany
azade.farshad@tum.de
[2] Munich Center for Machine Learning, Munich, Germany
[3] Stanford University, Stanford, USA
[4] Johns Hopkins University, Baltimore, USA

Abstract. Although the preservation of shape continuity and physiological anatomy is a natural assumption in the segmentation of medical images, it is often neglected by deep learning methods that mostly aim for the statistical modeling of input data as pixels rather than interconnected structures. In biological structures, however, organs are not separate entities; for example, in reality, a severed vessel is an indication of an underlying problem, but traditional segmentation models are not designed to strictly enforce the continuity of anatomy, potentially leading to inaccurate medical diagnoses. To address this issue, we propose a graph-based approach that enforces the continuity and connectivity of anatomical topology in medical images. Our method encodes the continuity of shapes as a graph constraint, ensuring that the network's predictions maintain this continuity. We evaluate our method on three public benchmarks of retinal vessel segmentation and one neuronal structure segmentation benchmark, showing significant improvements in connectivity metrics compared to previous works while getting better or on-par performance on segmentation metrics.

1 Introduction

Deep learning models for medical image analysis are mostly designed to prioritize the statistical modeling of shapes and textures over the morphology of the medical images. Even though the disruption of organs like vessels is often an indication of an underlying disorder, they are often neglected in favor of overall segmentation performance [1–3]. The importance of modeling structural integrity becomes more prominent for the diagnosis of sensitive organs, such as the eyes, which contain dense vasculature and pose an even greater challenge [4]. That is why such models are mostly evaluated and applied in analyzing fundus images. Although deep learning-based methods have improved the performance in retinal vessel segmentation [5–7], they often fail to preserve the continuity of shapes, which is crucial

Y. Yeganeh, G. Güvercin and R. Xiao—Equal Contribution.

© The Author(s), under exclusive license to Springer Nature Switzerland AG 2024
S.-A. Ahmadi and S. Pereira (Eds.): MICCAI 2023, LNCS 14373, pp. 3–13, 2024.
https://doi.org/10.1007/978-3-031-55088-1_1

for accurate medical analysis. To address this, we propose a novel graph-based approach that enforces shape continuity for image segmentation.

Previous studies have attempted to incorporate topological information into retinal vessel segmentation. One such approach is to design specific loss functions that penalize the disconnectedness of segmented retinal vessels. For example, Yan et al. [8] proposed joint segment-level and pixel-wise losses, which emphasize the thickness consistency of thin vessels. Shit et al. [9] developed the Centerline Dice (clDice) for tubular structures, a connectivity-aware similarity metric that improves segmentation results with more accurate connectivity information. Some studies have suggested topological objective functions to improve deep-learning models to produce results that have more similar topology as the ground truth, which help to reduce topological errors [10–12].

Other approaches incorporate architectures [13–15] that aim to maintain the spatial information in the image. The shape conservation can be improved by either introducing residual connections [16,17] or by using deformation-based architectures that rely on a prior mask with the same topological features as the ground truth. For example, Zhang et al. [11] proposed the TPSN, in which a deformation map generated from an encoder-decoder architecture transforms the template mask into the region of interest. Wyburd et al. [18] also developed TEDS-Net on a continuous diffeomorphic framework.

Structural graph neural networks provide another way of modeling the connectivity and continuity of biological structures. Shin et al. [19] integrated a graph attention network [20] into a convolutional neural network to exploit both local properties and global vessel structures. Li et al. [21] extended such models to segment hepatic vessels and modified the network to learn graphical connectivity from the ground truth directly. Yu et al. [22] maintained vessel topology by constructing edges and predicting links between different nodes generated from semantic information extracted from U-Net [23]. These works showed that graphs are able to represent the structures in an image, however, most of these works focused on the optimization of pixel-wise metrics.

Here, we introduce a direct edge construction strategy in which each vertex is connected to all 1-level neighboring vertices of its surrounding patches by undirected and unweighted edges. To capture the neighborhood in every direction, we include not only vertical and horizontal connections but also diagonal edges between the vertices. Therefore, our model is able to represent the underlying structures of an image in a granular grid form by using direct, unconstrained connections in the graph, which eliminates the necessity of geodesic vessel interpretation in previous work [19,22]. Moreover, we adapt the clDice to operate on the graphs, instead of the pixel values. This adapted loss function directly optimizes the connectivity of nodes in the graph, enforcing the preservation of the continuity in the graph.

Our proposed graph-based model is agnostic to prior CNN segmentation [24] and directly addresses the issue of preserving shape continuity in medical image segmentation, as opposed to other graph construction approaches in previous works, such as VGN [19]. Unlike previous approaches that rely on selecting the pixel with the highest probability in non-overlapping regions [19–22], our method

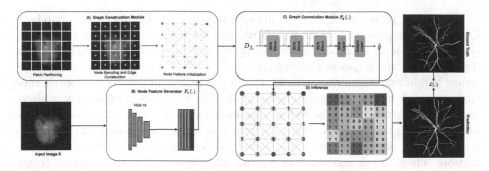

Fig. 1. The visual graph is constructed using patches and 1-level edge connections in (A), node features are generated in (B), graph features are extracted in (C) and combined with initial spatial features using skip connections, and final image segmentation is produced in (D) by extending node predictions into corresponding patches.

treats the patches in the image as a node for graph construction, resulting in a more efficient approach to represent the underlying structures.

Contrary to CNN models, in which the information from the input can affect the inference, by enforcing the relation of neighboring patches of the image via message passing, we inherently bound each part of the input to the neighboring patches affecting the final inference. Through comprehensive evaluation, we demonstrate that our approach significantly improves shape continuity metrics compared to the baselines while maintaining comparable or better performance on pixel-wise segmentation metrics.

2 Methodology

In this section, we present our continuity preservation network. First, we introduce the overview of the proposed framework (also see Fig. 1). Then, we elaborate on the components of our graph-based continuity-aware model.

2.1 Scope

We propose a graph-based architecture for our continuity-aware model since graphs enforce connectivity by design based on the intrinsic connections between the nodes in the graph. Our model comprises four general steps to predict a segmentation map from the input image. First, we construct a graph from the input image x (Fig. 1-A), then we embed the visual features obtained from the image using a pretrained feature extraction network into the graph nodes (Fig. 1-B). The graph embeddings are fed to a graph neural networks (GNNs) [25] to process the features by message passing through the graph (Fig. 1-C). Finally, the processed graph features are encoded using the segmentation framework to predict the segmentation map y.

The input image $x \in \mathbb{R}^{H \times W}$ is divided into non-overlapping $n \times n$ patches. Each patch, represented as an individual graph vertex v_i, connects to its corresponding 1-level neighboring vertices by the edges of a fully-connected graph denoted with the adjacency matrix $A \in \mathbb{R}^{H/n \times W/n}$, where H and N are the image height and width respectively. The graph representation comprises N vertices, with each vertex responsible for predicting its corresponding $n \times n$ sized patch. To generate initial node-level features for these vertices, we designed a node feature generator $F_c(.)$ that receives the input image, extracts its pixel-wise features, and maps them into node-level features as vertex feature initialization. Then, graph construction module $G(.)$ fuses all sampled graph nodes with their corresponding feature vectors to complete this graph construction process, where $g_{\{A, f_c\}} = G(x, F_c(x))$. The generated features hold geometrical information about the vertices enabling them to represent general structures of the organs like vessels; however, they do not enforce any constraints over their relations. To introduce topological information graph convolution module $F_g(.)$ extracts node-level graph features and combines them with their initial spatial features to make final class predictions for the nodes $\hat{y} \in \mathbb{R}^{N \times 2}$ given ground-truth labels $y \in \mathbb{R}^N$, where $\hat{y} = F_g(A, f_c)$.

Graph Construction. To construct the visual graph, we first define an image grid of non-overlapping, equally-sized, square $n \times n$ patches, where each patch serves as a vertex of the graph. This results in a vertex set $V = v_i{}_{i=1}^N$, where $N = H/n, \times, W/n$ in an input image $x \in \mathbb{R}^{H, \times, W}$, as shown in Fig. 1-A. Each vertex is descriptive of its corresponding patch and is initialized with feature vectors generated for each of those patches.

Node Feature Generation. Node feature generation is a crucial step in our architecture, as it maps the pixel-wise features to the corresponding vertex-level features. We follow the approach proposed by Maninis et al. [26], where a VGG-16 backbone [24] is used to extract the spatial features of a retinal fundus image $x \in \mathbb{R}^{H, \times, W}$. We obtain a multi-level feature representation $f \in \mathbb{R}^{H, \times, W, \times, 64}$ from the VGG backbone, capturing both local and global pixel-wise features. To generate the node-level features for our graph we use a 1×1 convolution with a max-pool layer of $n \times n$ window and a stride of n, where n is the patch size. This operator downsamples the pixel-wise feature maps to a resolution of $H/n \times W/n$, generating node-level feature maps $f_c \in \mathbb{R}^{H/n, \times, W/n, \times, 64}$. By applying this operator, we obtain a more compact and efficient feature representation for each vertex, which is important for training our segmentation model. Figure 1-B shows the node feature generation process.

Graph Convolution Module. We propose our graph convolution module $F_g(.)$, based on node classification with graph convolution operator [27], at the top of our grid-based visual graph of the nodes to learn feature representations encoding the relationship and neighborhood between the nodes. First, initial node features f_c are reformatted into node feature matrix $D_{f_c} \in \mathbb{R}^{N \times 64}$ for graph-based processing. Then, D_{f_c} and A are fed into our graph convolution module to generate graph features combined with initial node features by a skip connection, which results in final node feature matrix $D_{f_{g+c}} \in \mathbb{R}^{N \times 96}$. Class prediction for each node, to decide whether the patch of the node contains a vessel branch or not, is made by the last graph convolution layer by using those final node features. There are, in total, eleven graph convolution layers in the network, in which the input to layer i is concatenated with the output of layer $i + 2$ by local skip connections to be passed into the layer $i + 3$ as shown in Fig. 1-C. These skip connections enable the network to have easier gradient flow inside $F_g(.)$ and also between $F_g(.)$ and $F_c(.)$.

Objective Functions. Our graph-based approach utilizes two loss functions: 1) a conventional CE loss, and 2) clDice [9]. CE loss helps optimize the node classification accuracy, while clDice preserves the continuity of the vessel structures by measuring the centerline connectivity of the segmented and ground-truth vessels. However, since clDice is designed for spatial domain images, not for graph-based representations, we propose an alternation of clDice for graphs to project the output node predictions of the GCN module as well as the input vessel graphs into 2D space. These projections are then fed to the clDice to compute the loss.

Inference. Our inference module (shown in Fig. 1-D) is responsible for providing a smooth transition between node predictions and their pixel-wise correspondences. After the graph convolution module classifies each node as foreground or background, the class predictions are extended to $n \times n$ spatial areas that correspond to the nodes in the input image, resulting in a segmentation map in the spatial domain. Our approach differs depending on the size of the node patches. If we use 1×1 patches, each pixel is treated as a node, and the node predictions are directly accepted as pixel predictions for the segmentation map. However, if we use larger patch sizes such as 2×2, each node represents a region of pixels. In this case, we use nearest neighbor interpolation to map the class of each node to its corresponding 2×2 region. This allows us to produce a pixel-wise segmentation map even when the node patches are larger than a single pixel. This way, our mapping function serves as a transition from the graph domain to the spatial domain, which facilitates the interpretation of visual graphs in the image space.

3 Experiments and Results

This section provides a comprehensive evaluation of the proposed method for segmenting retinal vessels based on various metrics. The evaluation is carried out on the CHASE_DB1 [28], FIVES [29], DRIVE [30] and ISBI2012 [31] datasets. We compare our methodology with state-of-the-art methods, and perform an ablation study of graph construction components. The details of the datasets and the results on DRIVE and ISBI2012 are presented in the supplementary material.

Table 1. Qualitative results on FIVES [29] dataset

| Segmentation Map | GT | DRIU [26] | RU-UNet [32] | SCOPE (Ours) |

Table 2. Comparison to SOTA on CHASE_DB1 [28]. The results denoted by * are evaluated in the 5-fold cross validation setting.

Method	Pixel-Wise Metrics			Connectivity Metrics			
	Pre (%) ↑	Rec (%) ↑	Dice(%)↑	clDice (%) ↑	$error_{\beta_0}$ ↓	$error_{\beta_1}$ ↓	$error_\chi$ ↓
RU-UNet [32]	0.78	0.80	0.79	0.74	59.8	2.8	61.2
DRIU [26]	0.79	0.80	0.79	0.74	37.8	4.2	35.1
VGN [19]	0.59	**0.94**	0.73	0.78	71.9	4.4	69.5
FR-UNet [15]	0.76	0.88	0.80	0.73	61.0	2.8	64.4
SGL [7]	0.79	0.87	**0.82**	0.75	42.6	2.3	46.0
Graph Cuts Loss [33]	**0.81**	0.69	0.73	0.75	93.7	5.2	90.6
SCOPE (Ours)	0.75	0.86	0.80	**0.81**	**24.2**	**1.6**	**22.7**
Graph Cuts Loss* [33]	**0.83**	0.64	0.71	0.74	106.2	7.7	97.7
DconnNet* [34]	0.79	0.73	**0.81**	**0.83**	64.2	7.0	57.2
SCOPE* (Ours)	0.76	**0.83**	**0.81**	**0.83**	**22.0**	**1.7**	**21.0**

3.1 Experimental Setup

All values reported for the RU-UNet [32] and DRIU [26] experiments are based on our own implementation. We train the baseline models as well as ours on the FIVES [29] dataset for 100 epochs, using Adam optimizer with a batch size of 8 and weight decay of $1e^{-3}$. We set the learning rate to $1e^{-2}$ for RU-UNet [14] and $1e^{-3}$ for DRIU [26]. To address the domain shift problem and adapt the pretrained weights to the context of CHASE [28], we fine-tuned the baseline models that are pre-trained on FIVES [29]. We used the same hyperparameters for all datasets unless specified. The models were trained on CHASE for 200 epochs. To be consistent with the original paper, we used cross-entropy loss for both datasets during training. We downsample the images to 512×512 resolution for training and testing in both datasets. We evaluate our method using the same data and metrics as previous works. However, since DeconnNet [34] uses 5-fold cross validation, for fairness, we also apply the same data-splitting strategy to our proposed method.

Table 3. Comparison with previous work. We present the results of our methodology compared to previous work on the FIVES [29] dataset.

Class	Architecture	Pixel-Wise Metrics			Connectivity Metrics			
		Pre (%) ↑	Rec (%) ↑	Dice(%) ↑	clDice (%) ↑	$error_{\beta_0}$ ↓	$error_{\beta_1}$ ↓	$error_\chi$ ↓
AMD	RU-UNet [32]	0.88	0.93	0.90	0.87	72.8	4.3	71.1
	DRIU [26]	0.87	**0.95**	**0.91**	0.87	57.8	2.7	57.1
	SCOPE (Ours)	**0.90**	0.92	**0.91**	**0.94**	**6.7**	**2.6**	**7.9**
DR	RU-UNet [32]	0.85	0.92	0.89	0.84	89.5	4.1	83.8
	DRIU [26]	0.85	**0.94**	0.89	0.85	64.3	3.8	60.5
	SCOPE (Ours)	**0.91**	0.89	**0.90**	**0.92**	**10.6**	**3.0**	**14.0**
Glaucoma	RU-UNet [32]	0.84	0.90	0.87	0.83	87.6	5.6	83.2
	DRIU [26]	0.87	**0.92**	**0.89**	**0.85**	51.2	**3.4**	53.1
	SCOPE (Ours)	**0.91**	0.77	0.81	0.83	**5.4**	4.5	**11.9**
Normal	RU-UNet [32]	0.87	0.90	**0.88**	0.84	141.6	8.8	123.3
	DRIU [26]	0.83	**0.94**	**0.88**	0.84	138.9	6.4	124.8
	SCOPE (Ours)	**0.89**	0.87	**0.88**	**0.89**	**14.0**	**4.2**	**13.2**
Average	RU-UNet [32]	0.85	0.90	0.87	0.83	98.0	5.6	90.5
	DRIU [26]	0.85	**0.92**	**0.88**	0.84	73.8	4.5	70.4
	SCOPE (Ours)	**0.90**	0.85	0.85	**0.87**	**12.1**	**3.7**	**14.4**

For our proposed method, the node feature generator is trained individually on each dataset using the same hyper-parameters as the DRIU [26] baseline. We incorporate the node feature generator into our graph convolution module, and it is jointly trained with graph-integrated clDice loss [9]. We set the learning rate for training SCOPE on FIVES to $4e^{-4}$, and the weight decay to $5e^{-3}$. For CHASE [28], we adhere to the same configuration as in FIVES [29] except for the number of training epochs which is 200 due to the limited size of the dataset.

Evaluation Metrics. Given that the intended goal in this paper is to evaluate the continuity and topology of vessels rather than pure pixel-wise metrics, we use four main topological metrics in our results: (i) clDice [9] evaluating the topological continuity of tubular structures, (ii) β_0 (Betti zero) [10] that counts the number of connected components in a topological space, (iii) β_1 (Betti one) [10] indicating the number of independent loops or cycles in the space, and (iv) χ (Euler) characteristic [35] that is a topological invariant quantifying the connectivity and complexity anatomical structure by counting the number of its connected components, holes, and voids, we utilize them as $error_{\beta_0}$, $error_{\beta_1}$, and $error_\chi$ to compare ground truth with predicted results.

3.2 Results

Comparison of vessel segmentation models on CHASE dataset [28] using pixel-wise and connectivity metrics shown on Table 2, indicates that our model significantly outperforms other methods in continuity metrics and comparable pixel-wise metrics values, outperforming state-of-the-art architectures and two baseline models, a recent modification of RU-UNet [14], and DRIU [26].

Table 4. Sample Images with Ground-Truth and SCOPE Predictions on Fives Dataset

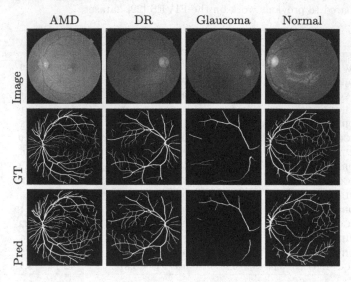

Table 5. Ablation study on different loss functions and patch sizes on FIVES [29].

Loss		Pixel-Wise Metrics			Connectivity Metrics			
		Pre (%) ↑	Rec (%) ↑	Dice(%) ↑	clDice (%) ↑	$error_{\beta_0}$ ↓	$error_{\beta_1}$ ↓	$error_\chi$ ↓
	CE Loss	0.89	0.77	0.81	0.78	52.6	4.4	53.6
	clDice	**0.90**	**0.85**	**0.85**	**0.87**	**12.1**	**3.7**	**14.4**
n	1×1	**0.90**	**0.85**	**0.85**	**0.87**	12.1	**3.7**	14.4
	2×2	0.72	0.83	0.76	0.83	**10.29**	4.2	**13.2**

We also validate our proposed method on the newly released FIVES dataset [29] in Table 3. To the best of our knowledge, no method has reported results on the FIVES [29] dataset. To this end, we compare our model against two competitive models based on the results obtained on CHASE [28], namely RU-UNet [32], and DRIU [26], as the baselines on FIVES [29]. The results in Table 3 show that our model significantly outperforms the two baselines in connectivity metrics $error_{\beta_0}$, $error_{\beta_1}$, $error_\chi$, and clDice, while achieving better or comparable results in pixel-wise metrics. This improvement in connectivity metrics indicates that our model enforces the preservation of topological information, while other methods do not strictly enforce it. Our qualitative results in Table 1 also support that where other baselines introduced discontinuity to the topology, our method preserves it. Note that since pixel-wise metrics are only sensitive to the pixels, they cannot represent connectivity. In addition, we provide the comparison of our models to Swin-UNETR [36] and Attention-UNET [37] in the supplementary material.

We conduct ablation studies Table 5 to examine the impact of the objective function and patch size on the segmentation performance. We observe that clDice

[9] significantly improves both pixel-wise metrics and connectivity metrics, as it is specifically designed for tubular structures. We also find that increasing the patch size leads to worse pixel-wise metrics Table 5, however, the 2×2 patch size achieves better betti0 and Euler than 1×1. This suggests that larger patch sizes specialized the model to capture global information about the vessel structures, rather than pixel-wise information. Depending on the domain of data, changing the patch size could potentially lead to the preservation of the continuity and topology, which are reflected by the connectivity metrics.

4 Conclusion

In conclusion, preserving shape continuity is critical in medical imaging as it can impact the accuracy of medical diagnoses. Unfortunately, traditional deep learning methods often neglect this aspect, leading to inaccurate predictions. In this work, we proposed a graph-based approach that enforces shape continuity in medical segmentation by encoding it as a graph constraint. Our method significantly improved shape continuity metrics compared to traditional methods while maintaining or improving segmentation performance. By enforcing shape continuity, our approach can potentially improve the accuracy of medical diagnoses, especially in structures such as vessels, where continuity is critical.

References

1. Farshad, A., Makarevich, A., Belagiannis, V., Navab, N.: MetaMedSeg: volumetric meta-learning for few-shot organ segmentation. In MICCAI Workshop on Domain Adaptation and Representation Transfer, pp. 45–55. Springer (2022). https://doi.org/10.1007/978-3-031-16852-9_5
2. Yeganeh, Y., Farshad, A., Weinberger, P., Ahmadi, S.-A., Adeli, E., Navab, N.: Transformers pay attention to convolutions leveraging emerging properties of ViTs by dual attention-image network. In: Proceedings of the IEEE/CVF International Conference on Computer Vision (2023)
3. Farshad, A., Yeganeh, Y., Navab, N.: Learning to learn in medical applications: a journey through optimization. In: Meta-Learning with Medical Imaging and Health Informatics Applications, pp. 3–25. Elsevier (2023)
4. Farshad, A., Yeganeh, Y., Gehlbach, P., Navab, N.: Y-Net: a spatiospectral dual-encoder network for medical image segmentation. In: International Conference on Medical Image Computing and Computer-Assisted Intervention, pp. 582–592. Springer, Cham (2022). https://doi.org/10.1007/978-3-031-16434-7_56
5. Guo, C., Szemenyei, M., Yi, Y., Wang, W., Chen, B., Fan, C.: SA-UNet: spatial attention u-net for retinal vessel segmentation. In: 2020 25th International Conference on Pattern Recognition (ICPR). IEEE (2021)
6. Kamran, S.A., Hossain, K.F., Tavakkoli, A., Zuckerbrod, S.L., Sanders, K.M., Baker, S.A.: RV-GAN: segmenting retinal vascular structure in fundus photographs using a novel multi-scale generative adversarial network. In: de Bruijne, M., et al. (eds.) MICCAI 2021. LNCS, vol. 12908, pp. 34–44. Springer, Cham (2021). https://doi.org/10.1007/978-3-030-87237-3_4

7. Zhou, Y., Yu, H., Shi, H.: Study group learning: improving retinal vessel segmentation trained with noisy labels. In: de Bruijne, M., et al. (eds.) MICCAI 2021. LNCS, vol. 12901, pp. 57–67. Springer, Cham (2021). https://doi.org/10.1007/978-3-030-87193-2_6

8. Yan, Z., Yang, X., Cheng, K.-T.: Joint segment-level and pixel-wise losses for deep learning based retinal vessel segmentation. IEEE Trans. Biomed. Eng. **65**(9), 1912–1923 (2018)

9. Shit, S., et al.: clDice-a novel topology-preserving loss function for tubular structure segmentation. In: Proceedings of CVPR, pp. 16560–16569 (2021)

10. Clough, J.R., Byrne, N., Oksuz, I., Zimmer, V.A., Schnabel, J.A., King, A.P.: A topological loss function for deep-learning based image segmentation using persistent homology. IEEE Trans. Pattern Anal. Mach. Intell. **44**(12), 8766–8778 (2020)

11. Zhang, H., Lui, L.M.: Topology-preserving segmentation network: a deep learning segmentation framework for connected component. arXiv (2022)

12. Mozafari, M., Bitarafan, A., Azampour, M.F., Farshad, A., Baghshah, M.S., Navab., N.: VISA-FSS: a volume-informed self supervised approach for few-shot 3D segmentation. In: International Conference on Medical Image Computing and Computer-Assisted Intervention, pp. 112–122. Springer, (2023). https://doi.org/10.1007/978-3-031-43895-0_11

13. Zhang, Z., Liu, Q., Wang, Y.: Road extraction by deep residual u-net. IEEE Geosci. Remote Sens. Lett. **15**(5), 749–753 (2018)

14. Alom, M.Z., Yakopcic, C., Hasan, M., Taha, T.M., Asari, V.K.: Recurrent residual U-Net for medical image segmentation. J. Med. Imaging **6**(1), 014006–014006 (2019)

15. Liu, W., et al.: Full-resolution network and dual-threshold iteration for retinal vessel and coronary angiograph segmentation. IEEE J. Biomed. Health Inform. **26**(9), 4623–4634 (2022)

16. Zhuang, J.: LadderNet: multi-path networks based on U-Net for medical image segmentation. arXiv (2018)

17. Li, L., Verma, M., Nakashima, Y., Nagahara, H., Kawasaki, R.: IterNet: retinal image segmentation utilizing structural redundancy in vessel networks. In: Proceedings of WACV, pp. 3656–3665 (2020)

18. Wyburd, M.K., Dinsdale, N.K., Namburete, A.I.L., Jenkinson, M.: TEDS-Net: enforcing diffeomorphisms in spatial transformers to guarantee topology preservation in segmentations. In: de Bruijne, M., et al. (eds.) MICCAI 2021. LNCS, vol. 12901, pp. 250–260. Springer, Cham (2021). https://doi.org/10.1007/978-3-030-87193-2_24

19. Shin, S.Y., Lee, S., Yun, I.D., Lee, K.M.: Deep vessel segmentation by learning graphical connectivity. Med. Image Anal. **58**, 101556 (2019)

20. Veličković, P., Cucurull, G., Casanova, A., Lio, P., Bengio, Y.: Graph attention networks. arXiv, Adriana Romero (2017)

21. Li, R., et al.: 3D graph-connectivity constrained network for hepatic vessel segmentation. IEEE J. Biomed. Health Inform. (2022)

22. Yu, H., Zhao, J., Zhang, L.: Vessel segmentation via link prediction of graph neural networks. In: Multiscale Multimodal Medical Imaging: Third International Workshop, MMMI 2022, pp. 34–43. Springer, Cham (2022). https://doi.org/10.1007/978-3-031-18814-5_4

23. Ronneberger, O., Fischer, P., Brox, T.: U-Net: convolutional networks for biomedical image segmentation. In: Medical Image Computing and Computer-Assisted Intervention-MICCAI 2015, pp. 234–241. Springer, Cham (2015). https://doi.org/10.1007/978-3-319-24574-4_28

24. Simonyan, K., Zisserman, A.: Very deep convolutional networks for large-scale image recognition. arXiv (2014)
25. Scarselli, F., Gori, M., Tsoi, A.C., Hagenbuchner, M., Monfardini, G.: The graph neural network model. IEEE Trans. Neural Netw. **20**(1), 61–80 (2008)
26. Maninis, K.-K., Pont-Tuset, J., Arbeláez, P., Van Gool, L.: Deep retinal image understanding. In: Ourselin, S., Joskowicz, L., Sabuncu, M.R., Unal, G., Wells, W. (eds.) MICCAI 2016. LNCS, vol. 9901, pp. 140–148. Springer, Cham (2016). https://doi.org/10.1007/978-3-319-46723-8_17
27. Kipf, T.N., Welling, M.: Semi-supervised classification with graph convolutional networks. arXiv (2016)
28. Fraz, M.M., et al.: An ensemble classification-based approach applied to retinal blood vessel segmentation. IEEE Trans. Biomed. Eng. **59**(9), 2538–2548 (2012)
29. Jin, K., et al.: Fives: a fundus image dataset for artificial intelligence based vessel segmentation. Sci. Data **9**(1), 475 (2022)
30. Staal, J., Abràmoff, M.D., Niemeijer, M., Viergever, M.A., Van Ginneken, B.: Ridge-based vessel segmentation in color images of the retina. IEEE Trans. Med. Imaging (2004)
31. Arganda-Carreras, I., et al.: Crowdsourcing the creation of image segmentation algorithms for connectomics. Front. Neuroanat. (2015)
32. Kerfoot, E., Clough, J., Oksuz, I., Lee, J., King, A.P., Schnabel, J.A.: Left-ventricle quantification using residual U-Net. In: Pop, M., et al. (eds.) STACOM 2018. LNCS, vol. 11395, pp. 371–380. Springer, Cham (2019). https://doi.org/10.1007/978-3-030-12029-0_40
33. Zheng, Z., Oda, M., Mori, K.: Graph cuts loss to boost model accuracy and generalizability for medical image segmentation. In: Proceedings of the ICCV, pp. 3304–3313 (2021)
34. Yang, Z., Farsiu, S.: Directional connectivity-based segmentation of medical images. In: Proceedings of the CVPR, pp. 11525–11535 (2023)
35. Beltramo, G., Andreeva, R., Giarratano, Y., Sarkar, R., Skraba, P.: Euler characteristic surfaces. arXiv, Miguel O Bernabeu (2021)
36. Hatamizadeh, A., Nath, V., Tang, Y., Yang, D., Roth, H.R., Xu, D.: Swin UNETR: swin transformers for semantic segmentation of brain tumors in MRI images. In: MICCAI Brainlesion Workshop. Springer, Cham (2021). https://doi.org/10.1007/978-3-031-08999-2_22
37. Oktay, O., et al.: Attention u-net: learning where to look for the pancreas. arXiv (2018)

Extended Graph Assessment Metrics for Regression and Weighted Graphs

Tamara T. Mueller[1(✉)], Sophie Starck[1], Leonhard F. Feiner[1,2],
Kyriaki-Margarita Bintsi[3], Daniel Rueckert[1,3], and Georgios Kaissis[1,2,4]

[1] Institute for AI in Medicine and Healthcare, Faculty of Informatics,
Technical University of Munich, Munich, Germany
tamara.mueller@tum.de
[2] Department of Diagnostic and Interventional Radiology, Faculty of Medicine,
Technical University of Munich, Munich, Germany
[3] BioMedIA, Department of Computing, Imperial College London, London, UK
[4] Institute for Machine Learning in Biomedical Imaging, Helmholtz-Zentrum Munich,
Munich, Germany

Abstract. When re-structuring patient cohorts into so-called population graphs, initially independent patients can be incorporated into one interconnected graph structure. This population graph can then be used for medical downstream tasks using graph neural networks (GNNs). The construction of a suitable graph structure is a challenging step in the learning pipeline that can have a severe impact on model performance. To this end, different graph assessment metrics have been introduced to evaluate graph structures. However, these metrics are limited to classification tasks and discrete adjacency matrices, only covering a small subset of real-world applications. In this work, we introduce extended graph assessment metrics (GAMs) for regression tasks and weighted graphs. We focus on two GAMs in particular: *homophily* and *cross-class neighbourhood similarity* (CCNS). We extend the notion of GAMs to more than one hop, define homophily for regression tasks, as well as continuous adjacency matrices, and propose a lightweight CCNS distance for discrete and continuous adjacency matrices. We show the correlation of these metrics with model performance on different medical population graphs and under different learning settings, using the TADPOLE and UKBB datasets[1](The source code can be found at https://github.com/tamaramueller/ExtendedGAMs).

Keywords: Graph neural networks · graph assessment metrics · medical population graphs

1 Introduction

The performance of graph neural networks can be highly dependent on the graph structure they are trained on [15,16]. To this end, several graph assessment metrics (GAMs) have been introduced to evaluate graph structures and shown strong correlations between specific graph structures and the performance of graph

S.-A. Ahmadi and S. Pereira (Eds.): MICCAI 2023, LNCS 14373, pp. 14–26, 2024.
https://doi.org/10.1007/978-3-031-55088-1_2

neural networks (GNNs) [14–16]. Especially in settings, where the graph structure is not provided by the dataset but needs to be constructed from the data, GAMs are the only way to assess the quality of the constructed graph. This is for example the case when utilising so-called population graphs on medical datasets. Recent works have furthermore shown that learning the graph structure in an end-to-end manner can improve performance on population graphs [9]. Some of these methods that learn the graph structure during model training operate with fully connected, weighted graphs, where all nodes are connected with each other and the tightness of the connection is determined by a learnable edge weight. This leads to a different representation of the graph, which does not fit the to-date formulations of GAMs. Furthermore, existing metrics are tailored to classification tasks and cannot be easily transformed for equally important regression tasks. The contributions of this work are the following: (1) We extend existing metrics to allow for an assessment of multi-hop neighbourhoods. (2) We introduce an extension of the homophily metric for regression tasks and continuous adjacency matrices and (3) define a cross-class neighbourhood similarity (CCNS) distance metric and extend CCNS to learning tasks that operate on continuous adjacency matrices. Finally, (4) we show these metrics' correlation to model performance on different medical and synthetic datasets. The metrics introduced in this work can find versatile applications in the area of graph deep learning in medical and non-medical settings since they strongly correlate with model performance and give insights into the graph structure in various learning settings.

2 Background and Related Work

2.1 Definition of Graphs

A discrete graph $G := (V, E)$ is defined by a set of n nodes V and a set of edges E, connecting pairs of nodes. The edges are unweighted and can be represented by an adjacency matrix $\mathbf{A}$ of shape $n \times n$, where $\mathbf{A}_{ij} = 1$ if and only if $e_{ij} \in E$ and 0 otherwise. A continuous/weighted graph $G_w := (V_w, E_w, \mathbf{W})$, assigns a (continuous) weight to every edge in E_w, summarised in the weight matrix $\mathbf{W}$. Continuous graphs are for example required in cases where the adjacency matrix is learned in an end-to-end manner and backpropagation through the adjacency matrix needs to be feasible. A neighbourhood $\mathcal{N}_v$ of a node v contains all direct neighbours of v and can be extended to k hops by $\mathcal{N}_v^{(k)}$. For this work, we assume familiarity with GNNs [3].

2.2 Homophily

Homophily is a frequently used metric to assess a graph structure that is correlated to GNN performance [15]. It quantifies how many neighbouring nodes share the same label [15] as the node of interest. There exist three different notions of homophily: edge homophily [10], node homophily [19], and class homophily [12,15]. Throughout this work, we use node homophily, sometimes omitting the term "node", only referring to "homophily".

Definition 1 (Node homophily). *Let $G := (V, E)$ be a graph with a set of node labels $Y := \{y_u; u \in V\}$ and $\mathcal{N}_v$ be the set of neighbouring nodes to node v. Then G has the following node homophily:*

$$h(G, Y) := \frac{1}{|V|} \sum_{v \in V} \frac{|\{u|u \in \mathcal{N}_v, Y_u = Y_v\}|}{|\mathcal{N}_v|}, \tag{1}$$

where $|\cdot|$ indicates the cardinality of a set.

A graph G with node labels Y is called *homophilous/homophilic* when $h(G,Y)$ is large (typically larger than 0.5) and *heterophilous/heterophilic* otherwise [10].

2.3   Cross-Class Neighbourhood Similarity

Ma et al. [16] introduce a metric to assess the graph structure for graph deep learning, called cross-class neighbourhood similarity (CCNS). This metric indicates how similar the neighbourhoods of nodes with the same labels are over the whole graph – irrespective of the labels of the neighbouring nodes.

Definition 2 (Cross-class neighbourhood similarity). *Let $G = (V, E)$, $\mathcal{N}_v$, and Y be defined as above. Let C be the set of node label classes, and $\mathcal{V}_c$ the set of nodes of class c. Then the CCNS of two classes c and c' is defined as follows:*

$$\text{CCNS}(c, c') = \frac{1}{|\mathcal{V}_c||\mathcal{V}_{c'}|} \sum_{u \in \mathcal{V}, v \in \mathcal{V}'} \text{cossim}(d(u), d(v)). \tag{2}$$

$d(v)$ is the histogram of a node v's neighbours' labels and $\text{cossim}(\cdot, \cdot)$ the cosine similarity.

3   Extended Graph Metrics

In this section, we introduce our main contributions by defining new extended GAMs for regression tasks and continuous adjacency matrices. We propose (1) a unidimensional version of CCNS which we call *CCNS distance*, which is easier to evaluate than the whole original CCNS matrix, (2) an extension of existing metrics to k-hops, (3) GAMs for continuous adjacency matrices, and (4) homophily for regression tasks.

3.1   CCNS Distance

The CCNS of a dataset with n classes is an $n \times n$ matrix, which can be large and cumbersome to evaluate. The most desirable CCNS for graph learning has high intra-class and low inter-class values, indicating similar neighbourhoods for the same class and different neighbourhoods between classes. We propose to collapse the CCNS matrix into a single value by evaluating the L_1 distance between the CCNS and the identity matrix, which we term *CCNS distance*.

Definition 3 (CCNS distance). *Let $G = (V, E)$, C, CCNS be defined as above. Then the CCNS distance of G is defined as follows:*

$$D_{\text{CCNS}} := \frac{1}{n} \sum \|\text{CCNS} - \mathbb{I}\|_1, \tag{3}$$

where $\mathbb{I}$ indicates the identity matrix and $\|\cdot\|_1$ the L_1 norm.

We note that the *CCNS distance* is best at low values and that we do not define CCNS for regression tasks, since it requires the existence of class labels.

3.2   *K*-Hop Metrics

Most GAMs only evaluate direct neighbourhoods. However, GNNs can apply the message-passing scheme to more hops, including more hops in the node feature embedding. We therefore propose to extend homophily and CCNS on unweighted graphs to k-hop neighbourhoods. An extension of the metrics on weighted graphs is more challenging since the edge weights impact the k-hop metrics. The formal definitions for k-hop homophily and CCNS for unweighted graphs can be found in the Appendix. We here exchange the notion of $\mathcal{N}_v$ with the specific k-hop neighbourhood $\mathcal{N}_v^{(k)}$ of interest.

3.3   Metrics for Continuous Adjacency Matrices

Several graph learning settings, such as [6,9], utilise a continuous graph structure. In order to allow for an evaluation of those graphs, we here define GAMs on the weight matrix $\mathbf{W}$ instead of the binary adjacency matrix $\mathbf{A}$.

Definition 4 (Homophily for continuous adjacency matrices). *Let $G_w = (V_w, E_w, \mathbf{W})$, be a weighted graph defined as above with a continuous adjacency matrix. Then the 1-hop node homophily of G_w is defined as follows:*

$$\text{HCont}(G_w, Y) := \frac{1}{|V|} \sum_{v \in V} \left(\frac{\sum_{u \in \mathcal{N}_v | y_u = y_v} w_{uv}}{\sum_{u \in \mathcal{N}_v} w_{uv}} \right), \tag{4}$$

where w_{uv} is the weight of the edge from u to v.

Definition 5 (CCNS for continuous adjacency matrices). *Let $G_w = (V_w, E_w, \mathbf{W})$, C, $\text{cossim}(\cdot, \cdot)$ be defined as above. Then, the CCNS for weighted graphs is defined as follows:*

$$\text{CCNS}_{cont}(c, c') := \frac{1}{|\mathcal{V}_c||\mathcal{V}_{c'}|} \sum_{u \in \mathcal{V}, v \in \mathcal{V}'} \text{cossim}(d_c(u), d_c(v)), \tag{5}$$

where $d_c(u)$ is the histogram considering the edge weights of the continuous adjacency matrix of the respective classes instead of the count of neighbours. The *CCNS distance* for continuous adjacency matrices can be evaluated as above.

3.4 Homophily for Regression

Homophily is only defined for node classification tasks, which strictly limits its application to a subset of use cases. However, many relevant graph learning tasks perform a downstream node regression, such as age regression [2,21]. We here define homophily for node regression tasks. Since homophily is a metric ranging from 0 to 1, we contain this range for regression tasks by normalising the labels between 0 and 1 prior to metric evaluation. We subtract the average node label distance from 1 to ensure the same range as homophily for classification.

Definition 6 (Homophily for regression). *Let* $G = (V, E)$ *and* $\mathcal{N}_v^k$ *be defined as above and* Y *be the vector of node labels, which is normalised between* 0 *and* 1. *Then the k-hop homophily for regression is defined as follows:*

$$\text{HReg}^{(k)}(G, Y) := 1 - \left(\frac{1}{|V|} \sum_{v \in V} \left(\frac{1}{|\mathcal{N}_v^{(k)}|} \sum_{n \in \mathcal{N}_v^{(k)}} \|y_v - y_n\|_1 \right) \right), \quad (6)$$

where $\|\cdot\|_1$ *indicates the* L_1 *norm.*

Definition 7 (Homophily for continuous adjacency matrices for regression). *Let* $G_w = (V_w, E_w, \mathbf{W})$, Y, *and* N_v *be defined as above and the task be a regression task, then the homophily of* G *is defined as follows:*

$$\text{HReg}(G, Y) := 1 - \left(\frac{1}{|V|} \sum_{v \in V} \left(\frac{\sum_{n \in \mathcal{N}_v} w_{nv} \|y_v - y_n\|_1}{\sum_{n \in \mathcal{N}_v} w_{nv}} \right) \right), \quad (7)$$

where w_{nv} *is the weight of the edge from* n *to* v *and* $\|\cdot\|_1$ *the* L_1 *norm.*

3.5 Metric Evaluation

In general, we recommend the evaluation of GAMs separately on the training, validation, and test set. We believe this to be an important evaluation step since the metrics can differ significantly between the different sub-graphs, given that the graph structure in only optimised on the training set.

4 Experiments and Results

We evaluate our metrics on several datasets with different graph learning techniques: We (1) assess benchmark classification datasets using a standard learning pipeline, and (2) medical population graphs for regression and classification that learn the adjacency matrix end-to-end. All experiments are performed in a transductive learning setting using graph convolutional networks (GCNs) [11]. In order to evaluate all introduced GAMs, we specifically perform experiments on two task settings: classification and regression, and under two graph learning settings: one using a discrete adjacency matrix and one using a continuous one.

4.1 Datasets

In order to evaluate the above-defined GAMs, we perform node-level prediction experiments with GNNs on different datasets. We evaluate $\{1, 2, 3\}$-hop homophily and CCNS distance on the benchmark citation datasets Cora, Cite-Seer, and PubMed [24], Computers and Photos, and Coauthors CS datasets [20]. All of these datasets are classification tasks. We use k-layer GCNs and compare performance to a multi-layer perceptron (MLP).

Furthermore, we evaluate the introduced metrics on two different medical population graph datasets, as well as two synthetic datasets. The baseline results for these datasets can be found in Appendix Table 4. We generate **synthetic datasets** for classification and regression to analyse the metrics in a controllable setting. As a real-world medical classification dataset, we use **TADPOLE** [17], a neuro-imaging dataset which has been frequently used for graph learning on population graphs [6,9,18]. For a regression population graph, we perform brain age prediction on 6 406 subjects of the UK BioBank [22] (**UKBB**). We use 22 clinical and 68 imaging features extracted from the subjects' magnetic resonance imaging (MRI) brain scans, following the approach in [5]. In both medical population graphs, each subject is represented by one node and similar subjects are either connected following the k-nearest neighbours approach, like in [9] or starting without any edges.

4.2 GNN Training

Prior to this work, the homophily metric only existed for an evaluation on discrete adjacency matrices. In this work, we extend this metric to continuous adjacency matrices. In order to evaluate the metrics for both, discrete and continuous adjacency matrices, we use two different graph learning methods: (a) *dDGM* and (b) *cDGM* from [9]. DGM stands for "differentiable graph module", referring to the fact that both methods learn the adjacency matrix in an end-to-end manner. cDGM hereby uses a continuous adjacency matrix, allowing us to evaluate the metrics introduced specifically for this setting. dDGM uses a discrete adjacency matrix by sampling the edges using the Gumbel-Top-K trick [8]. Both methods are similar in terms of model training and performance, allowing us to compare the newly introduced metrics to the existing homophily metric in the dDGM setting.

4.3 Results

(1) Benchmark Classification Datasets. The results on the benchmark datasets are summarised in Table 1. We can see that the k-hop metric values can differ greatly between the different hops for some datasets while staying more constant for others. This gives an interesting insight into the graph structure over several hops. We believe an evaluation of neighbourhoods in graph learning to be more insightful if the number of hops in the GNN matches the number of hops considered in the graph metric. Interestingly, the performance of k-hop GCNs

Table 1. K-hop graph metrics of benchmark node classification datasets. Cl.: number of classes, Nodes: number of nodes

Dataset	Nodes	Cl.	Node homophily ↑			D_{CCNS} ↓		
			1-hop	2-hop	3-hop	1-hop	2-hop	3-hop
Cora	1,433	7	0.825 ± 0.29	0.775 ± 0.26	0.663 ± 0.29	0.075	0.138	0.229
CiteSeer	3,703	6	0.706 ± 0.40	0.754 ± 0.28	0.712 ± 0.29	0.124	0.166	0.196
PubMed	19,717	3	0.792 ± 0.35	0.761 ± 0.26	0.687 ± 0.26	0.173	0.281	0.363
Computers	13,752	10	0.785 ± 0.26	0.569 ± 0.27	0.303 ± 0.20	0.080	0.275	0.697
Photo	7,650	8	0.837 ± 0.25	0.660 ± 0.30	0.447 ± 0.28	0.072	0.210	0.429
Coauthor CS	18,333	15	0.832 ± 0.24	0.698 ± 0.25	0.520 ± 0.25	0.043	0.110	0.237

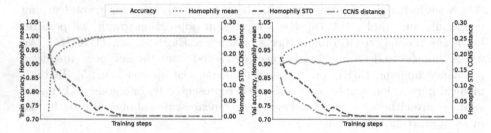

Fig. 1. Development of graph metrics on TADPOLE over training using **cDGM**; left: train set; right: validation set

did not align with the k-hop metric values on the specific datasets. We summarise these results in Appendix Table 3. One possible reason for this might be that, e.g., the 3-hop metrics assess the 1, 2, and 3-hop neighbourhood at once, not just the outer ring of neighbours. Another reason for this discrepancy might be that homophily and CCNS do not perfectly predict GNN performance. Furthermore, different graph convolutions have been shown to be affected differently by low-homophily graphs [25]. We believe this to be an interesting direction to further investigate GAMs for GNNs.

(2) Population Graph Experiments. Table 2 shows the dDGM and cDGM results of the population graph datasets. We can see that in some settings, such as the classification tasks on the synthetic dataset using dDGM, the homophily varies greatly between training and test set. This can be an indication of over-fitting on the training set since the graph structure is optimised for the training nodes only and might not generalise well to the whole graph.

Since we here use graph learning methods which adapt the graph structure during model training, also the graph metrics change over training. Figure 1 shows the development of the accuracy as well as the mean and standard deviation of the 1-hop homophily and CCNS distance, evaluated on the train (left) and validation set (right). We can see that for both sets, the homophily increases with the accuracy, while the standard deviation (STD) of the homophily decreases and the CCNS distance decreases with increasing performance. However, the GAMs

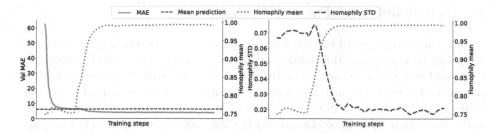

Fig. 2. Development of metrics on UKBB dataset using **cDGM** on validation set

Table 2. cDGM and dDGM results on the population graph datasets. We report the test scores averaged over 5 random seeds and 1-hop homophily and CCNS distance of one final model each. We do not report CCNS distance on regression datasets, since it is not defined for regression tasks.

Method	Dataset	Task	Test score	1-hop node homophily ↑		1-hop D_{CCNS} ↓	
				train	test	train	test
cDGM	Synthetic 1k	c	0.7900 ± 0.08	1.0000 ± 0.00	1.0000 ± 0.00	0.0000	0.0000
		r	0.0112 ± 0.01	0.9993 ± 0.00	0.9991 ± 0.00	–	–
	Synthetic 2k	c	0.8620 ± 0.03	1.0000 ± 0.00	1.0000 ± 0.00	0.0000	0.0000
		r	0.0173 ± 0.00	0.8787 ± 0.06	0.8828 ± 0.05	–	–
	Tadpole	c	0.9333 ± 0.01	1.0000 ± 0.00	0.9781 ± 0.09	0.0000	0.0314
	UKBB	r	4.0775 ± 0.23	0.8310 ± 0.06	0.8306 ± 0.07	–	–
dDGM	Synthetic 1k	c	0.8080 ± 0.04	0.6250 ± 0.42	0.1150 ± 0.32	0.4483	0.4577
		r	0.0262 ± 0.00	0.7865 ± 016	0.8472 ± 0.15	–	–
	Synthetic 2k	c	0.7170 ± 0.06	0.6884 ± 0.40	0.0950 ± 0.29	0.4115	0.4171
		r	0.0119 ± 0.00	0.8347 ± 0.13	0.8295 ± 0.13	–	–
	Tadpole	c	0.9614 ± 0.01	0.9297 ± 0.18	0.8801 ± 0.31	0.1045	0.0546
	UKBB	r	3.9067 ± 0.04	0.8941 ± 0.13	0.9114 ± 0.12	–	–

align more accurately with the training accuracy (left), showing that the method optimised the graph structure on the training set. The validation accuracy does not improve much in this example, while the validation GAMs still converge similarly to the ones evaluated on the train set (left). Figure 2 shows the mean (left) and STD (right) of the validation regression homophily HReg on the UKBB dataset with continuous adjacency matrices (using cDGM) and the corresponding change in validation mean absolute error (MAE). Again, homophily raises when the validation MAE decreases and the STD of the homophily decreases in parallel. On the left, the dotted grey line indicates the MAE of a mean prediction on the dataset. We can see that the mean regression homophily HReg raises once the validation MAE drops below the error of a mean prediction. We here only visualise a subset of all performed experiments, but we observe the same trends for all settings. From these experiments, we conclude that the here introduced GAMs show a strong correlation with model performance and can be used to assess generated graph structures that are used for graph deep learning.

5   Conclusion and Future Work

In this work, we extended two frequently used graph assessment metrics (GAMs) for graph deep learning, that allow to evaluate the graph structure in regression tasks and continuous adjacency matrices. For datasets that do not come with a pre-defined graph structure, like population graphs, the assessment of the graph structure is crucial for quality checks on the learning pipeline. Node homophily and cross-class neighbourhood similarity (CCNS) are commonly used GAMs that allow us to evaluate how similar the neighbourhoods in a graph are. However, these metrics are only defined for discrete adjacency matrices and classification tasks. This only covers a small portion of graph deep learning tasks. Several graph learning tasks target node regression [1,2,21]. Furthermore, recent graph learning methods have shown that end-to-end learning of the adjacency matrix is beneficial over statically creating the graph structure prior to learning [9]. These methods do not operate on a static binary adjacency matrix, but use weighted continuous graphs, which is not considered by most current GAMs. In order to overcome these limitations, we extend the definition of node homophily to regression tasks and both node homophily and CCNS to continuous adjacency matrices. We formulate these metrics and evaluate them on different synthetic and real-world medical datasets and show their strong correlation with model performance. We believe these metrics to be essential tools for investigating the performance of GNNs, especially in the setting of population graphs or similar settings that require explicit graph construction.

Our definition of the CCNS distance D_{CCNS} uses the L_1-norm to determine the distance between the node labels in order to weight each inter-class connection equally. However, the L_1-norm is only one of many norms that could be used here. Given the strong correlation of our definition of D_{CCNS}, we show that the usage of the L_1-norm is a sensible choice. We also see an extension of the metrics for weighted graphs to multiple hops as promising next steps towards better graph assessment for GNNs.

There exist additional GAMs, such as normalised total variation and normalised smoothness value [13], neighbourhood entropy and centre-neighbour similarity [23], and aggregations similarity score and diversification distinguishability [7] that have been shown to correlate with GNN performance. An extension of these metrics to regression tasks and weighted graphs would be interesting to investigate in future works. All implementations of the here introduced metrics are differentiable. This allows for seamless integration in the learning pipeline, e.g. as loss components, which could be a highly promising application to improve GNN performance by optimising for specific graph properties.

Acknowledgements. TM and SS were supported by the ERC (Deep4MI - 884622). This work has been conducted under the UK Biobank applications 87802 and 18545. SS has furthermore been supported by BMBF and the NextGenerationEU of the European Union. The data collection and sharing of the TADPOLE dataset was funded by the Alzheimers Disease Neuroimaging Initiative (ADNI) (National Institutes of

Health Grant U01 AG024904) and DOD ADNI (Department of Defense award number W81XWH-12-2-0012).

A   Further Information on Extended Graph Assessment Metrics

A.1 K-Hop metrics

We here formally define k-hop node homophily and k-hop CCNs.

Definition 8 (k-hop node homophily). *A graph $G := (V, E)$ with the set of node labels $Y := \{y_u; u \in V\}$ has the following k-hop node homophily:*

$$h^{(k)}(G, Y) := \frac{1}{|V|} \sum_{v \in V} \frac{\left|\{u | u \in \mathcal{N}_v^{(k)}, y_u = y_v\}\right|}{|\mathcal{N}_v^{(k)}|}, \tag{8}$$

where $\mathcal{N}_v^{(k)}$ is the set of nodes in the k-hop neighbourhood of v.

Definition 9 (k-hop CCNS). *A graph $G := (V, E)$ with the set of node labels C has the following k-hop CCNS for two classes c and c':*

$$\mathrm{CCNS}(c, c') = \frac{1}{|\mathcal{V}_c||\mathcal{V}_{c'}|} \sum_{u \in \mathcal{V}, v \in \mathcal{V}'} \mathrm{cossim}(d^{(k)}(u), d^{(k)}(v)), \tag{9}$$

where $d^{(k)}(v)$ indicates the empirical histogram of the labels of the k-hop neighbours of node v and $\mathrm{cossim}(\cdot, \cdot)$ the cosine similarity.

Table 3 summarises model performances of $\{1, 2, 3\}$-hop GCNs on the different benchmark datasets and the corresponding MLP performance on the node features only. We can see that even though 3-hop homophily of the datasets Computers and Photo is very low, the GCNs with 3 hops perform best on these datasets. This does not align with our initial intuition about these metrics and we believe this finding to be interesting to investigate further.

A.2   Node-wise metrics

The k-hop homophily for regression can also defined for every node individually and then combined in the full homophily over the entire graph as defined in the main part of this work.

Definition 10 (Homophily for regression). *Let $G = (V, E)$ and $\mathcal{N}_v^k$ be defined as above and Y be the vector of node labels, which is normalised between 0 and 1. Then the k-hop homophily of a node $v \in V_c$ in a node regression task is defined as the mean label distance between the node v and all its neighbours.*

$$\mathrm{HReg}_v^{(k)} := 1 - \left(\frac{1}{|\mathcal{N}_v^{(k)}|} \sum_{n \in \mathcal{N}_v^{(k)}} \|y_v - y_n\|\right), \tag{10}$$

where $|\cdot|$ is the cardinality of a set and $\|x\|$ the absolute value of x.

Table 3. Graph metrics of benchmark node classification datasets with corresponding performances of an MLP and 1,2, and 3-hop GCNs, reported in accuracy in %. Nodes: number of nodes, Cl.: number of classes in the dataset.

Dataset	Nodes	Cl.	Node homophily			MLP	GCN		
			1-hop	2-hop	3-hop		1-hop	2-hop	3-hop
Cora	1 433	7	0.825 ± 0.29	0.775 ± 0.26	0.663 ± 0.29	60.41	76.33	**81.70**	78.90
Citeseer	3 703	6	0.706 ± 0.40	0.754 ± 0.28	0.712 ± 0.29	61.19	71.20	**72.10**	67.10
Pubmed	19 717	3	0.792 ± 0.35	0.761 ± 0.26	0.687 ± 0.26	74.00	76.60	**79.10**	77.70
Computers	13 752	10	0.785 ± 0.26	0.569 ± 0.27	0.303 ± 0.20	79.35	39.27	67.56	**83.13**
Photo	7 650	8	0.837 ± 0.25	0.660 ± 0.30	0.447 ± 0.28	82.09	48.10	82.88	**88.37**
Coauthor CS	18 333	15	0.832 ± 0.24	0.698 ± 0.25	0.520 ± 0.25	88.93	**93.13**	89.31	92.09

The k-hop homophily for regression of the whole graph G can then be extracted as follows:

$$
\mathrm{HReg}_G^{(k)} := 1 - \left(\frac{1}{|V|} \sum_{v \in V} \mathrm{HReg}_v^{(k)} \right)
$$

$$
= 1 - \left(\frac{1}{|V|} \sum_{v \in V} \left(\frac{1}{|\mathcal{N}_v^{(k)}|} \sum_{n \in \mathcal{N}_v^{(k)}} \|y_v - y_n\| \right) \right). \tag{11}
$$

B   Experiments

In this section, we give more details on training parameters and setups of the experiments performed in this work.

B.1   Synthetic dataset

The synthetic datasets are generated using *sklearn* [4]. Each dataset consist of either 1 000 or 2 000 nodes, with 50 node features of which 5 are informative. For all experiments on the synthetic dataset, we utilise early stopping and the initial graph structure is generated using the k-nearest neighbours approach with 5 neighbours and the Euclidean distance.

B.2   TADPOLE dataset

We use the same TADPOLE dataset as in [9], which consists of 564 subjects. The task is the classification of Alzheimer's disease, mild cognitive impairment and control normal. For all experiments on the TADPOLE dataset, we use early stopping and generate the initial graph structure using the k-nearest neighbours approach and the Euclidean distance.

B.3   UKBB dataset

The dDGM experiments on the UKBB dataset are performed using no initial graph structure since this resulted in better model performance. For the cDGM experiments we use the k-nearest neighbours approach with x neighbours. We utilised no early stopping and the Euclidean distance for the graph construction for the cDGM experiments.

B.4   Baseline results

Table 4 summarises the baseline results on the population graph datasets using a random forest and the implementation from `sklearn` [4].

Table 4. Baseline results using random forests on the different datasets. For classification tasks, we report accuracy in % and for regression MAE. We report the mean and standard deviation of a 5-fold cross-validation.

Dataset	Nr. nodes	Task	Test Score
Synthetic	1000	Binary classification	78.00 ± 0.07
		Regression	0.0529 ± 0.01
	2000	Binary classification	88.10 ± 0.02
		Regression	0.0081 ± 0.00
Tadpole	564	Classification	94.15 ± 0.01
UKBB	6406	Regression	4.2644 ± 0.05

References

1. Berrone, S., Della Santa, F., Mastropietro, A., Pieraccini, S., Vaccarino, F.: Graph-informed neural networks for regressions on graph-structured data. Mathematics **10**(5) (2022). https://doi.org/10.3390/math10050786, https://www.mdpi.com/2227-7390/10/5/786
2. Bintsi, K.M., Baltatzis, V., Potamias, R.A., Hammers, A., Rueckert, D.: Multimodal brain age estimation using interpretable adaptive population-graph learning. arXiv preprint arXiv:2307.04639 (2023)
3. Bronstein, M.M., Bruna, J., LeCun, Y., Szlam, A., Vandergheynst, P.: Geometric deep learning: going beyond Euclidean data. IEEE Signal Process. Mag. **34**(4), 18–42 (2017)
4. Buitinck, L., et al.: API design for machine learning software: experiences from the scikit-learn project. In: ECML PKDD Workshop: Languages for Data Mining and Machine Learning, pp. 108–122 (2013)
5. Cole, J.H.: Multimodality neuroimaging brain-age in UK biobank: relationship to biomedical, lifestyle, and cognitive factors. Neurobiol. Aging **92**, 34–42 (2020)
6. Cosmo, L., Kazi, A., Ahmadi, S.-A., Navab, N., Bronstein, M.: Latent-graph learning for disease prediction. In: Martel, A.L., et al. (eds.) MICCAI 2020. LNCS, vol. 12262, pp. 643–653. Springer, Cham (2020). https://doi.org/10.1007/978-3-030-59713-9_62

7. Elam, J.S., et al.: The human connectome project: a retrospective. Neuroimage **244**, 118543 (2021)
8. Jiang, B., Zhang, Z., Lin, D., Tang, J., Luo, B.: Semi-supervised learning with graph learning-convolutional networks. In: Proceedings of the IEEE/CVF Conference on Computer Vision and Pattern Recognition, pp. 11313–11320 (2019)
9. Kazi, A., Cosmo, L., Ahmadi, S.A., Navab, N., Bronstein, M.M.: Differentiable graph module (DGM) for graph convolutional networks. IEEE Trans. Pattern Anal. Mach. Intell. **45**(2), 1606–1617 (2022)
10. Kim, D., Oh, A.: How to find your friendly neighborhood: graph attention design with self-supervision. arXiv preprint arXiv:2204.04879 (2022)
11. Kipf, T.N., Welling, M.: Semi-supervised classification with graph convolutional networks. arXiv preprint arXiv:1609.02907 (2016)
12. Lim, D., Li, X., Hohne, F., Lim, S.N.: New benchmarks for learning on non-homophilous graphs. arXiv preprint arXiv:2104.01404 (2021)
13. Lu, S., Zhu, Z., Gorriz, J.M., Wang, S.H., Zhang, Y.D.: NAGNN: classification of COVID-19 based on neighboring aware representation from deep graph neural network. Int. J. Intell. Syst. **37**(2), 1572–1598 (2022)
14. Luan, S., Hua, C., Lu, Q., Zhu, J., Chang, X.W., Precup, D.: When do we need GNN for node classification? arXiv preprint arXiv:2210.16979 (2022)
15. Luan, S., et al.: Is heterophily a real nightmare for graph neural networks to do node classification? arXiv preprint arXiv:2109.05641 (2021)
16. Ma, Y., Liu, X., Shah, N., Tang, J.: Is homophily a necessity for graph neural networks? arXiv preprint arXiv:2106.06134 (2021)
17. Mueller, S.G., et al.: The Alzheimer's disease neuroimaging initiative. Neuroimaging Clin. **15**(4), 869–877 (2005)
18. Parisot, S., et al.: Spectral graph convolutions for population-based disease prediction. In: Descoteaux, M., Maier-Hein, L., Franz, A., Jannin, P., Collins, D.L., Duchesne, S. (eds.) MICCAI 2017. LNCS, vol. 10435, pp. 177–185. Springer, Cham (2017). https://doi.org/10.1007/978-3-319-66179-7_21
19. Pei, H., Wei, B., Chang, K.C.C., Lei, Y., Yang, B.: Geom-GCN: geometric graph convolutional networks. arXiv preprint arXiv:2002.05287 (2020)
20. Shchur, O., Mumme, M., Bojchevski, A., Günnemann, S.: Pitfalls of graph neural network evaluation. arXiv preprint arXiv:1811.05868 (2018)
21. Stankeviciute, K., Azevedo, T., Campbell, A., Bethlehem, R., Lio, P.: Population graph GNNs for brain age prediction. In: Proceedings of the ICML, vol. 202 (2020)
22. Sudlow, C., et al.: UK biobank: an open access resource for identifying the causes of a wide range of complex diseases of middle and old age. PLoS Med. **12**(3), e1001779 (2015)
23. Xie, Y., Li, S., Yang, C., Wong, R.C.W., Han, J.: When do GNNs work: understanding and improving neighborhood aggregation. In: IJCAI'20: Proceedings of the Twenty-Ninth International Joint Conference on Artificial Intelligence, {IJCAI} 2020, vol. 2020 (2020)
24. Yang, Z., Cohen, W., Salakhudinov, R.: Revisiting semi-supervised learning with graph embeddings. In: International Conference on Machine Learning, pp. 40–48. PMLR (2016)
25. Zhu, J., Yan, Y., Zhao, L., Heimann, M., Akoglu, L., Koutra, D.: Beyond homophily in graph neural networks: current limitations and effective designs. Adv. Neural. Inf. Process. Syst. **33**, 7793–7804 (2020)

Multi-head Graph Convolutional Network for Structural Connectome Classification

Anees Kazi[1,2(✉)], Jocelyn Mora[1], Bruce Fischl[1,2], Adrian V. Dalca[1,2,3], and Iman Aganj[1,2]

[1] Athinoula A. Martinos Center for Biomedical Imaging, Radiology Department, Massachusetts General Hospital, Boston, USA
akazi1@mgh.harvard.edu
[2] Radiology Department, Harvard Medical School, Boston, USA
[3] CSAIL, Massachusetts Institute of Technology, Cambridge, USA

Abstract. We tackle classification based on brain connectivity derived from diffusion magnetic resonance images. We propose a machine-learning model inspired by graph convolutional networks (GCNs), which takes a brain-connectivity input graph and processes the data separately through a parallel GCN mechanism with multiple heads. The proposed network is a simple design that employs different heads involving graph convolutions focused on edges and nodes, thoroughly capturing representations from the input data. To test the ability of our model to extract complementary and representative features from brain connectivity data, we chose the task of sex classification. This quantifies the degree to which the connectome varies depending on the sex, which is important for improving our understanding of health and disease in both sexes. We show experiments on two publicly available datasets: PREVENT-AD (347 subjects) and OASIS3 (771 subjects). The proposed model demonstrates the highest performance compared to the existing machine-learning algorithms we tested, including classical methods and (graph and non-graph) deep learning. We provide a detailed analysis of each component of our model.

1 Introduction

Structural connections between brain regions constitute complex brain networks known as the *connectome*. Brain networks are represented by graphs, where each brain region is a node, with edges representing the connections between regions and the edge weight reflecting the strength of the connection. The resulting graph provides a detailed map of brain structural connectivity and can be used to study the organization of brain networks and how they relate to cognitive function and behavior. Structural connectome graphs created from diffusion MRI (dMRI) have been used to study a wide range of neurological and psychiatric disorders, including Alzheimer's disease (AD) [3,10,25], schizophrenia [16,26], and autism spectrum disorders (ASD) [24], as well as to understand normal brain development and aging [8].

In this line, sex classification using structural brain connectivity has been an important problem [6,28]. Clinically, understanding sex differences in brain

S.-A. Ahmadi and S. Pereira (Eds.): MICCAI 2023, LNCS 14373, pp. 27–36, 2024.
https://doi.org/10.1007/978-3-031-55088-1_3

connectivity patterns can provide insights into the neurobiology of neurological and psychiatric disorders that have different prevalence rates and symptoms
between males and females [11]. For example, simple thresholding of the structural brain connectivity has shown different sub-brain networks in males versus
females [14,15]. Some studies have found that males and females with ASD [6]
and conduct disorder [30] have different patterns of structural brain connectivity, which may help to explain differences in the symptomatology of the disorder
between the sexes. Therefore, predicting sex based on the structural connectome
may help to identify potential biomarkers or risk factors for these disorders and
to develop more personalized and effective treatments.

The task of classification using the structural connectome involves analyzing high-dimensional and complex data, which can be challenging for traditional
statistical approaches. Graph neural networks (GNNs), particularly graph convolutional networks (GCNs) [31], provide a powerful and flexible framework for
analyzing brain connectivity graphs. GCNs can learn from the complex interrelationships between nodes and edges, capturing both local and global patterns in the graph structure. This makes GCNs well-suited for classification and
prediction based on structural brain connectivity and for identifying the most
predictive brain regions and connections. Furthermore, GCNs can be trained
on large datasets, increasing their generalizability and applicability to different
populations and contexts.

GCNs can leverage the rich structural information in the connectome to
make accurate predictions. They do so by performing iterative message-passing
between neighboring nodes in the graph, using learnable functions to aggregate
and transform information from neighboring nodes, and updating the features
of each node based on the aggregated information. This allows GCNs to capture
the complex relationships between brain regions and their connections and make
predictions based on this information. A GCN-based model is promising for
analyzing structural connectome and has shown great potential for improving
our understanding of neurological and psychiatric disorders, as well as normal
brain development and aging.

Several state-of-the-art methods have shown the application of GCNs in sex
classification from structural and functional brain networks, achieving high classification accuracy compared to existing machine-learning (ML) methods. In a
study combining the GCN model and the long short-term memory (LSTM) network to categorize the functional connectivity of demented and healthy patients
[29], to enhance the disease classification, gender and age predictions were added
to a regularization task. Further, the Siamese GCN has been proposed for metric
learning in the context of sex classification [19]. GCNs are also combined with
recurrent neural networks to predict sex on temporal fMRI brain graphs [17].
The spectral GCN has been employed for the region-of-interest identification in
functional connectivity graphs for sex classification as well [5].

In this paper, we propose a simple yet efficient GCN-based multi-head model
capable of differentiating sex using structural brain connectivity. We propose a
new design architecture that involves multiple parallel graph/non-graph based

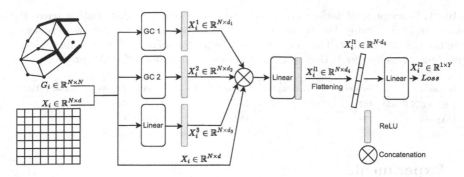

Fig. 1. End-to-end pipeline of the proposed model. GC stands for Graph Convolution. The thickness of the edges in G_i shows the weights on the edges. d, d_1, d_2, d_3, d_4 are the output dimensions at each layer. $l1$ and $l2$ are the outputs of the two linear layers.

operations to collect information from all fronts. Through experiments on two public databases, we show that the proposed model outperforms conventional ML and non-graph-based deep-learning (DL) methods. We continue with a description of the proposed method, experiments, discussion, and conclusion.

2   Methods

Let the dataset be $D = (D_1, D_2, ..., D_M)$ with M subjects. The ith subject is represented as $D_i \in (G_i, X_i)$, i.e., with a brain connectivity graph $G_i \in \mathbb{R}^{N \times N}$ with N nodes, and the corresponding feature matrix $X_i \in \mathbb{R}^{N \times d}$ representing the features for the nodes. The task is to classify each subject D_i into Y classes. We define a model f_θ as:

$$y_i = f_\theta(G_i, X_i), \tag{1}$$

where θ is the set of learnable parameters. The proposed model f_θ consists of four branches collecting complementary information from the same input setup. We employ a combination of GCNConv layers [18], linear layers, and a skip connection. The GCNConv layers capture low-level features of the graph, while the linear layers learn complex, non-linear relationships between features and make the final classification decision. Lastly, the skip connection helps to address the specific problem of over-smoothing, which may occur when applying graph convolutions. GCNConv is based on the graph Laplacian matrix and uses a simple convolutional operation to propagate information between neighboring nodes in the graph. It can be mathematically defined as $X_i^l = S^{-\frac{1}{2}} G_i S^{\frac{1}{2}} X_i \Theta_l$ where S is the diagonal degree matrix, $S_{jj} = \sum_{k=1}^{N} G_{jk}$, and Θ_l is the set of parameters for the lth branch. The linear layers in the model are defined as $X_i^l = \sigma(\Theta_l X_i + b)$, where σ is the non-linearity (ReLU) function and b is the bias. The motivation for using a combination of GCNConv layers with different embedding sizes is that each one transforms the data into a different space from the same input, hence collecting varied information. The end-to-end pipeline is shown in Fig. 1. We use the weighted cross-entropy loss to train the model.

Table 1. Description of dataset size (number of available scans), distribution across the classes, and partitioning. Due to missing demographic data, nine subjects were removed from the OASIS3 dataset. The female ratio is the portion of scans from female subjects, providing a baseline prediction accuracy for a constant (always female) predictor.

Name	Subjects	Total samples	Samples-10%	10%	Male	Female	Female ratio
PREVENT–AD	347	789	710	79	199	511	72%
OASIS3	771	1294	1164	121	515	649	56%

3   Experiments

We thoroughly analyzed our multi-head GCN model via the task of sex classification with various experiments on two public databases (Table 1). We set aside 10% of our data. The rest of the data was used for developing and fine-tuneing our models through cross-validation. Here, we first provide details on the datasets, pre-processing and implementation. We then show baseline experiments, followed by a comparison with the state-of-the-art DL-based methods. Further, we show ablation tests for various learning techniques used. Lastly, we show results on the 10% held-out data to check the model's generalizability.

3.1   Datasets

Pre-symptomatic Evaluation of Experimental or Novel Treatments for Alzheimer's Disease (PREVENT-AD) [21] is a publicly available dataset that aims to provide a comprehensive set of data on individuals who are at risk for developing AD (https://prevent-alzheimer.net). The database contains neuroimaging studies such as MRI (including dMRI) and PET scans, a range of demographic, clinical, cognitive, and genetic data, as well as data on lifestyle factors such as diet and exercise. The dataset comprises 347 subjects, some with multiple (longitudinal) dMRI scans, totaling 789 dMRI scans.

Open Access Series of Imaging Studies, the Third Release (OASIS3) [20] is a longitudinal neuroimaging, clinical, and cognitive dataset for normal aging and AD, provided freely to researchers worldwide (http://www.oasis-brains.org). The OASIS3 dataset contains MRI scans (including dMRI), cognitive assessments, demographic information, and clinical diagnoses for subjects, including healthy controls, individuals with MCI, and AD patients. We used 1294 brain scans from 771 subjects.

3.2   Pre-processing

We used FreeSurfer [9] to process the databases (additionally the longitudinal processing pipeline [23] for PREVENT-AD). We then ran the FreeSurfer diffusion processing pipeline and propagated the 85 automatically segmented cortical and subcortical regions from the structural to the diffusion space. These 85

Table 2. Classification results (mean accuracy) using conventional ML methods. **Bold** and red denote the best and the runner-up, respectively.

Model Type/Dataset	PREVENT-AD	OASIS3
Tree Coarse Tree	74.1	60.1
Logistic Regression	54.5	52.6
Naive Bayes (Kernel)	62.0	55.1
SVM (Quadratic)	**84.1**	**72.3**
KNN (Weighted KNN)	77.5	63.6
Ensemble (Boosted KNN)	78.3	68.0
Ensemble (Subspace Discriminant)	82.8	64.9
Ensemble (Subspace KNN)	74.2	59.5
Ensemble (RUSBoosted Trees)	72.5	67.3
Neural Network (Wide)	83.2	72.0

regions act as the nodes in our graph setup. Next, we used our public toolbox (http://www.nitrc.org/projects/csaodf-hough) to reconstruct the diffusion orientation distribution function in constant solid angle [2], run Hough-transform global probabilistic tractography [1] to generate 10,000 fibers per subject, compute symmetric structural connectivity matrices, and augment the matrices with indirect connections [4]. Once we had all the graphs G_i, we performed a population-level normalization on edge weights. For node features, we used the volume, apparent diffusion coefficient, and fractional anisotropy obtained for each region, as well as the row in G_i representing the connectivity to the rest of the brain. Therefore, for each subject we obtained $G_i \in \mathbb{R}^{85 \times 85}$ and corresponding $X_i \in \mathbb{R}^{85 \times 88}$.

Implementation Details. All the experiments were run via 10-fold cross-validation with the same folds across methods and experiments. The data was split into 10 folds based on subjects (rather than scans). For model robustness, we added zero-mean random normal noise with a standard deviation of 0.01 to the training samples. All the experiments were run on a Linux machine with 512 GB of RAM, an Intel (R) Xeon (R) Gold 6256 CPU @ 3.60 GHz, and an NVIDIA RTX A6000 (48 GB) graphics processing unit. For a fair comparison, we chose the number of layers for comparative methods such that the total numbers of parameters for GCNConv (2453), DGCNN (4653), Graphconv (4653), ResGatedGraphConv (RGGC) (9128), and GINConv (2683) were similar to the proposed method (5073). In our experiments, the values of d_1, d_2, d_3, and d_4 were 25, 20, 5, and 2, respectively. We kept 10% of the data aside from both datasets so as not to heuristically fit the model to the entire data, and tested the model at the end on the unseen data. All the comparative methods are selected from PyTorch geometric [12].

Baselines. Before testing our model, we applied conventional ML algorithms to determine the baseline performance for the two datasets. We used the Statistical and Machine Learning Toolbox of MATLAB with default parameters. Apart from basic classification methods such as decision trees (Coarse Tree), Logistic Regression, and Kernel-based Naive Bayes, we also tested the Support Vector Machine (SVM) and K-Nearest Neighbors. Table 2 presents the performance of these models on both datasets. The results suggest that the SVM (Quadratic) and Neural Network perform best among all the methods.

Table 3. Classification results (accuracy mean ± StD) using DL methods, with and without data augmentation (all with skip connections, but various respective pooling strategies).

	With Augmentation		Without Augmentation	
Model/Dataset	PREVENT-AD	OASIS3	PREVENT-AD	OASIS3
MLP	77.3 ± 6.5	75.1 ± 4.1	77.3 ± 6.6	75.1 ± 4.0
DGCNN[27]	76.8 ± 5.5	75.5 ± 3.8	76.9 ± 4.8	74.4 ± 4.4
Graphconv[22]	79.8 ± 1.3	75.2 ± 4.2	80.9 ± 6.5	75.3 ± 2.9
RGGC[7]	80.8 ± 7.2	74.8 ± 4.2	80.0 ± 7.4	75.3 ± 4.3
GINConv[13]	80.8 ± 4.2	74.3 ± 3.8	80.8 ± 3.5	74.1 ± 4.4
GCNConv[18]	85.2 ± 5.8	81.8 ± 5.2	85.9 ± 5.3	81.5 ± 4.4
Proposed	**90.6 ± 6.8**	**88.6 ± 4.1**	**89.5 ± 6.1**	**87.8 ± 6.6**

3.3   Results and Discussion

Comparative Methods. Table 3 shows the performance of various DL models with and without data augmentation. We show results on augmenting the node features with zero-mean uniform noise with $StD = 0.01$. The table includes the mean and standard deviation of accuracy (across folds) for each model and dataset. The proposed model, with data augmentation, achieves the highest accuracy for both datasets, with GCNConv yielding the second best performance. Overall, augmentation improved the performance and robustness of our model.

Ablation Tests. Table 4 shows the accuracy for each model and pooling technique, suggesting that the choice of pooling technique can significantly impact the performance of the models. We use flattening for aggregating the information from the graph nodes, which can be seen to perform best among pooling techniques. Flattening keeps information from the entire graph, i.e. all the nodes and their corresponding learned representation, whereas mean and max pooling smoothes and leaves out the information, respectively. This table highlights the importance of considering different pooling techniques when selecting and evaluating DL models. Table 5 shows the ablation results on adding a skip connection from the raw input to the final linear layer, suggesting an (insignificant) improvement by the skip connection in most cases. Table 3 also shows the ablation on

Table 4. Comparison of the performance (accuracy mean ± StD) of several GNN models with three different pooling techniques (no augmentation or skip connections).

Dataset	Model/Pooling	Max pool	Mean pool	Flattening
PREVENT-AD	DGCNN[27]	77.9 ± 6.1	76.9 ± 4.8	77.9 ± 6.1
	Graphconv[22]	80.6 ± 5.6	80.8 ± 6.5	80.9 ± 5.8
	RGGC[7]	80.6 ± 5.9	80.0 ± 7.4	81.3 ± 5.0
	GINConv[13]	79.9 ± 6.0	80.8 ± 3.5	81.8 ± 5.0
	GCNConv[18]	**85.6 ± 6.1**	85.9 ± 5.3	88.3 ± 5.5
	Proposed	83.6 ± 4.7	**87.5 ± 5.6**	**90.5 ± 5.3**
OASIS3	DGCNN[27]	73.8 ± 5.2	74.7 ± 3.9	74.9 ± 3.7
	Graphconv[22]	75.2 ± 3.8	75.0 ± 3.5	75.6 ± 3.6
	RGGC[7]	73.0 ± 3.9	76.0 ± 3.8	75.2 ± 4.6
	GINConv[13]	73.1 ± 3.9	73.4 ± 5.2	73.5 ± 4.2
	GCNConv[18]	**83.5 ± 4.1**	82.5 ± 4.1	83.5 ± 4.2
	Proposed	82.7 ± 4.1	**82.5 ± 4.1**	**86.1 ± 4.5**

Table 5. Experimental results (accuracy mean ± StD) with and without skip connection (all with flattening and no augmentation).

	PREVENT-AD		OASIS3	
Model/Skip	with skip	without skip	with skip	without skip
MLP	**74.7 ± 3.6**	74.6 ± 4.8	**74.7 ± 4.5**	73.9 ± 4.5
DGCNN	**79.7 ± 5.1**	77.9 ± 6.1	**75.3 ± 4.1**	74.9 ± 3.7
Graphconv	**82.1 ± 4.6**	80.9 ± 5.8	75.2 ± 3.7	**75.6 ± 3.6**
RGGG	80.3 ± 4.9	**81.3 ± 5.0**	74.7 ± 3.6	**75.2 ± 4.6**
GINConv	81.2 ± 4.3	**81.8 ± 5.0**	73.2 ± 3.3	**73.5 ± 4.2**
GCNConv	**88.5 ± 6.0**	88.3 ± 5.5	**87.0 ± 5.6**	83.5 ± 4.2
Proposed	**90.7 ± 6.3**	89.03 ± 5.4	**87.4 ± 6.6**	86.1 ± 4.5

augmentation. It can be observed that the results of the proposed method for Prevent-AD dataset is different despite the same setup, this is due to the randomness in initialization of two separate experiments.

Results on Held-Out Data. Finally, we tested our model on never-before-seen data. Table 6 shows results on 10% of the data that was kept aside before the model development and training. In this experiment, we took the pre-trained model and evaluated the classification accuracy on the new data, which reveals how the model would translate to relatively new data. The flattening technique is applied to the proposed method, whereas maxpooling is applied to comparative methods. It can be observed that the proposed model still performed best for PREVENT-AD and almost tied with DGCNN for OASIS3, showing its superiority with respect to out-of-sample performance.

Table 6. Classification accuracy of 10% held-out (never before seen) data.

Dataset/Method	MLP	DGCNN	Graphconv	RGGG	GINConv	GCNConv	Proposed
PREVENT-AD	65.6	73.1	69.9	72.0	54.8	75.3	**78.5**
OASIS3	69.4	**95.9**	88.4	90.9	90.1	89.3	95.0

4 Conclusion

In this paper, we proposed a simple yet effective model capable of capturing complementary information from brain connectivity graphs, which we evaluated in the context of sex classification. The configuration of input data, the initialization of neighborhood information as node features, the combination of GCNConv layers, linear layers, and a skip connection, and eventually the flattening of node features helped to learn better representations of each subject's graph. We have shown that our model outperforms competing techniques on two publicly available datasets, while also ablating several components (augmentation, pooling technique, skip connection). Our results on held-out data further help to measure the model's robustness toward unseen data. In terms of network complexity and size, the proposed model is average-sized, as mentioned in the implementation details. Future work includes the addition of interpretability to the models to find the brain subnetworks responsible for the sex difference, integration of functional and structural connectivity, and evaluation of disease and age prediction. A further step would be to try different graph convolution mechanisms, such as those based on residual connections or gated attention graph convolutions.

Clinical Translation. The proposed method, which takes advantage of GCNs, can be extended from sex classification to clinical prediction and stratification. For instance, it can be used as a biomarker for diagnosis, prognosis, progression/conversion prediction, and treatment effectiveness assessment.

Acknowledgments. Support for this research was provided by the National Institutes of Health (NIH), specifically the National Institute on Aging (NIA; RF1AG068261). Additional support was provided in part by the BRAIN Initiative Cell Census Network grant U01MH117023, the National Institute for Biomedical Imaging and Bioengineering (P41EB015896, R01EB023281, R01EB006758, R21EB018907, R01EB019956, P41EB030006), the NIA (R56AG064027, R01AG064027, R01AG008122, R01AG 016495, R01AG070988), the National Institute of Mental Health (R01MH121885, RF1MH123195), the National Institute for Neurological Disorders and Stroke (R01NS 0525851, R21NS072652, R01NS070963, R01NS083534, U01NS086625, U24NS10059103, R01NS105820), the NIH Blueprint for Neuroscience Research (U01MH093765), part of the multi-institutional Human Connectome Project, and the Michael J. Fox Foundation for Parkinson's Research (MJFF-021226). Computational resources were provided through the Massachusetts Life Sciences Center.
B. Fischl has a financial interest in CorticoMetrics, a company whose medical pursuits focus on brain imaging and measurement technologies. His interests were reviewed and

are managed by Massachusetts General Hospital and Mass General Brigham per their conflict-of-interest policies.

References

1. Aganj, I., et al.: A Hough transform global probabilistic approach to multiple-subject diffusion MRI tractography. Med. Image Anal. **15**(4), 414–425 (2011)
2. Aganj, I., Lenglet, C., Sapiro, G., Yacoub, E., Ugurbil, K., Harel, N.: Reconstruction of the orientation distribution function in single-and multiple-shell q-ball imaging within constant solid angle. Magn. Resonan. Med. **64**(2), 554–566 (2010)
3. Aganj, I., Mora, J., Frau-Pascual, A., Fischl, B., Initiative, A.D.N., et al.: Exploratory correlation of the human structural connectome with non-MRI variables in Alzheimer's disease. Alzheimer's Dement.: Diagn. Assess. Dis. Monit. (2023)
4. Aganj, I., Prasad, G., Srinivasan, P., Yendiki, A., Thompson, P.M., Fischl, B.: Structural brain network augmentation via Kirchhoff's laws. In: Joint Annual Meeting of ISMRM-ESMRMB, vol. 22, p. 2665 (2014). http://nmr.mgh.harvard.edu/~iman/ConductanceModel_ISMRM14_iman.pdf
5. Arslan, S., Ktena, S.I., Glocker, B., Rueckert, D.: Graph saliency maps through spectral convolutional networks: application to sex classification with brain connectivity. In: Graphs in Biomedical Image Analysis and Integrating Medical Imaging and Non-Imaging Modalities: Second International Workshop, GRAIL 2018 and First International Workshop, Beyond MIC 2018, Held in Conjunction with MICCAI 2018, Granada, Spain, September 20, 2018, Proceedings 2, pp. 3–13 (2018)
6. Beacher, F.D., et al.: Autism attenuates sex differences in brain structure: a combined voxel-based morphometry and diffusion tensor imaging study. Am. J. Neuroradiol. **33**(1), 83–89 (2012)
7. Bresson, X., Laurent, T.: Residual gated graph convnets. arXiv preprint arXiv:1711.07553 (2017)
8. Dennis, E.L., et al.: Development of brain structural connectivity between ages 12 and 30: a 4-tesla diffusion imaging study in 439 adolescents and adults. Neuroimage **64**, 671–684 (2013)
9. Fischl, B.: Freesurfer. Neuroimage **62**(2), 774–781 (2012)
10. Frau-Pascual, A., et al.: Conductance-based structural brain connectivity in aging and dementia. Brain Connect. **11**(7), 566–583 (2021)
11. Gur, R.E., Gur, R.C.: Sex differences in brain and behavior in adolescence: findings from the philadelphia neurodevelopmental cohort. N & B Reviews
12. He, Y., Zhang, X., Huang, J., Rozemberczki, B., Cucuringu, M., Reinert, G.: Pytorch geometric signed directed: a software package on graph neural networks for signed and directed graphs. arXiv preprint arXiv:2202.10793 (2022)
13. Hu, W., et al.: Strategies for pre-training graph neural networks. arXiv preprint arXiv:1905.12265 (2019)
14. Ingalhalikar, M., et al.: Sex differences in the structural connectome of the human brain. Proc. Natl. Acad. Sci. **111**(2), 823–828 (2014)
15. Jahanshad, N., et al.: Sex differences in the human connectome: 4-tesla high angular resolution diffusion imaging (hardi) tractography in 234 young adult twins. In: 2011 IEEE International Symposium on Biomedical Imaging: From Nano to Macro, pp. 939–943. IEEE (2011)
16. Karlsgodt, K.H., Sun, D., Cannon, T.D.: Structural and functional brain abnormalities in schizophrenia. Curr. Direct. Psychol. Sci. **19**(4), 226–231 (2010)

17. Kazi, A., et al.: DG-GRU: dynamic graph based gated recurrent unit for age and gender prediction using brain imaging. In: Medical Imaging 2022: Computer-Aided Diagnosis, vol. 12033, pp. 277–281. SPIE (2022)
18. Kipf, T.N., Welling, M.: Semi-supervised classification with graph convolutional networks. arXiv preprint arXiv:1609.02907 (2016)
19. Ktena, S.I., et al.: Metric learning with spectral graph convolutions on brain connectivity networks. Neuroimage **169**, 431–442 (2018)
20. LaMontagne, P.J., et al.: Oasis-3: longitudinal neuroimaging, clinical, and cognitive dataset for normal aging and Alzheimer disease. MedRxiv, pp. 2019–12 (2019)
21. Leoutsakos, J.M., Gross, A., Jones, R., Albert, M., Breitner, J.: 'Alzheimer's progression score': development of a biomarker summary outcome for ad prevention trials. The J. Prevent. Alzheimer's Disease **3**(4), 229 (2016)
22. Morris, C., et al.: Weisfeiler and leman go neural: higher-order graph neural networks. In: Proceedings of the AAAI Conference on Artificial Intelligence, vol. 33, pp. 4602–4609
23. Reuter, M., Schmansky, N.J., Rosas, H.D., Fischl, B.: Within-subject template estimation for unbiased longitudinal image analysis. Neuroimage **61**(4), 1402–1418 (2012)
24. Tolan, E., Isik, Z.: Graph theory based classification of brain connectivity network for autism spectrum disorder. In: Rojas, I., Ortuño, F. (eds.) IWBBIO 2018. LNCS, vol. 10813, pp. 520–530. Springer, Cham (2018). https://doi.org/10.1007/978-3-319-78723-7_45
25. Wang, J., et al.: Alterations in brain network topology and structural-functional connectome coupling relate to cognitive impairment. Front. Aging Neurosci. **10**, 404 (2018)
26. Wang, Y.M., et al.: Altered grey matter volume and white matter integrity in individuals with high Schizo-obsessive traits, high schizotypal traits and obsessive-compulsive symptoms. Asian J. Psychiatry **52**, 102096 (2020)
27. Wang, Y., Sun, Y., Liu, Z., Sarma, S.E., Bronstein, M.M., Solomon, J.M.: Dynamic graph CNN for learning on point clouds. ACM Trans. Graphics (tog)
28. Williamson, J., et al.: Sex differences in brain functional connectivity of hippocampus in mild cognitive impairment. Front. Aging Neurosci. (2022)
29. Xing, X., et al.: Dynamic spectral graph convolution networks with assistant task training for early MCI diagnosis. In: Shen, D., et al. (eds.) MICCAI 2019. LNCS, vol. 11767, pp. 639–646. Springer, Cham (2019). https://doi.org/10.1007/978-3-030-32251-9_70
30. Zhang, J., et al.: Sex differences of uncinate fasciculus structural connectivity in individuals with conduct disorder. BioMed Res. Int. (2014)
31. Zhang, S., Tong, H., Xu, J., Maciejewski, R.: Graph convolutional networks: a comprehensive review. Comput. Soc. Networks **6**(1), 1–23 (2019)

Tertiary Lymphoid Structures Generation Through Graph-Based Diffusion

Manuel Madeira[✉] [iD], Dorina Thanou[iD], and Pascal Frossard[iD]

École Polytechnique Fédérale de Lausanne (EPFL), Lausanne, Switzerland
manuel.madeira@epfl.ch

Abstract. Graph-based representation approaches have been proven to be successful in the analysis of biomedical data due to their capability of capturing intricate dependencies between biological entities, such as the spatial organization of different cell types in a tumor tissue. However, to further enhance our understanding of the underlying governing biological mechanisms, it is important to accurately capture the actual distributions of such complex data. Graph-based deep generative models are specifically tailored to accomplish that. In this work, we leverage state-of-the-art graph-based diffusion models to generate biologically meaningful cell-graphs. In particular, we show that the adopted graph diffusion model is able to accurately learn the distribution of cells in terms of their tertiary lymphoid structures (TLS) content, a well-established biomarker for evaluating the cancer progression in oncology research. Additionally, we further illustrate the utility of the learned generative models for data augmentation in a TLS classification task. To the best of our knowledge, this is the first work that leverages the power of graph diffusion models in generating meaningful biological cell structures.

Keywords: Deep generative models · Graph-based diffusion · Tertiary lymphoid structures

1 Introduction

Biomedical applications have largely benefited from graph-based approaches as a powerful framework for complex interactions modelling. The representation and analysis of the relationships between biologically-relevant entities has been exploited at different levels, ranging from more abstract dependencies, such as metabolic or gene regulatory networks [5,9,17], to completely observable relations, such as the spatial interactions of cells in digital pathology (DP) settings [1,12,13].

Under these settings, we are typically interested in tasks such as the understanding of metabolic or genetic interactions, patient diagnosis, or response prediction. A key factor to perform satisfactorily on those problems is the inference over previously unseen graphs and, to accomplish so, we need to correctly capture the underlying governing biological mechanisms. In particular, graph generation

S.-A. Ahmadi and S. Pereira (Eds.): MICCAI 2023, LNCS 14373, pp. 37–53, 2024.
https://doi.org/10.1007/978-3-031-55088-1_4

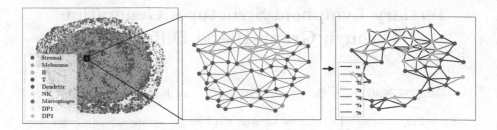

Fig. 1. Cell-graphs extracted from a global cell-graph, computed from a cancerous tissue sample imaged through multiplexed immunofluorescence with several cell type markers. *Left:* Cell-graph obtained from a whole-slide image. *Center:* High TLS content subgraph extracted from the cell-graph on the left. It is distinctly composed of a B-cell nucleus enveloped by T-cells. *Right:* Subgraph used for the TLS embeddings computation, following the procedure described in [23]. Only the edges between B- and T-cells are kept. If the vertices of the edge are of the same cell type, it is classified as α; otherwise, as γ_k, with k consisting of the number of B-cells neighboring the B-cell that is a vertex of the edge.

models seek to model the distribution of the actual graphs they are trained on. Thus, this type of approach can play a pivotal role for a better understanding of graph distributions, as well as on the generation of biologically plausible graph instances.

Diffusion models [11, 25] have recently emerged as the state-of-the-art approach for deep generative modelling by combining several desirable properties such as training stability, excellent sample quality, easy model scaling, and good distribution coverage. Meanwhile, several graph-based diffusion schemes have been proposed through the adaptation of either score-based generative modelling [14, 20] or of discrete denoising diffusion probabilistic models [2, 8, 27] to graphs. Adapting these to biological applications amenable to graph-based modelling holds great promise for further improvement in biological graphs mimicking.

In this work, we leverage on previously proposed graph-based diffusion approaches and extend their application to the cell-graph representation of biological tissues [7, 12]. Such graphs consider physical proximity between cells in tissue slides as a proxy to their interaction, assigning cells to nodes and drawing edges between adjacent cells, as depicted in Fig. 1. We focus specifically on the simple, yet biologically meaningful, tertiary lymphoid structures (TLS). TLS are highly structured biological entities composed of B-cell clusters surrounded by supporting T-cells, typically found in ectopic sites of chronic inflammation [21, 23]. These structures are correlated to a longer disease-free survival in cancer [6, 10, 16, 19, 23]. We then build on DiGress [27], a state-of-the-art graph generative model for the setting at hand (i.e., small graphs with categorical node and edge features), and extend it to the generation of cell-graphs with high and low TLS content.

We show that the trained generative model can indeed capture the underlying distribution by evaluating the similarity between the training graph population and a generated graph population in terms of TLS content [23]. To further illustrate the pertinence of the model to address the complexity of this task, we compare its performance to a non deep learning graph-based baseline. Furthermore, we also showcase the utility of generative models for data augmentation in a scarce data regime. We consider a binary classification task, where a Graph Convolutional Network (GCN) is used to classify input graphs as having high or low TLS content. We use the synthetically generated data to augment the training set of the GCN. We demonstrate that the generated graphs are sufficiently faithful to the real data distribution, leading to improved performance on the downstream classification task.

To the best of our knowledge, this work consists of the first graph-based generative approach for cell-graphs. It opens new promising venues in modelling distributions of structured tissues at a cell level, which can be a significant first step towards building more effective machine learning methods in digital pathology or personalised oncology.

2   Background

In this section, we start by introducing how graphs can be used to model biological entities and their relations. Then, we describe the relevance of generative models in the biomedical realm. We also briefly review graph-based diffusion methods.

Modelling Biomedical Structures with Graphs. In DP, tissue slides are digitized into whole-slide images (WSIs). The predominant applications of deep learning in DP rely on the extraction of image-level representations from those WSIs to a wide variety of tasks, ranging from object recognition, such as slide segmentation or structure detection, to higher level problems, such as cancer grading or survival prediction [3,24]. However, these approaches are typically limited in several aspects. WSIs are typically huge, requiring the tiling of the original image into smaller sized patches to be fed into an image based deep learning approach. This limitation imposes a trade-off on the context per patch that can be considered. Moreover, image based deep learning approaches model pixelwise relations, thus lacking efficient representations of biological entities and their relations. This fact also leads to a more convoluted interpretability of the obtained models. In contrast, entity-graphs based approaches have been shown to evade these limitations [1,12], yielding promising results both in predictive performance and interpretability [28]. Such graphs are composed of biological entities as nodes and relations between entities as edges [7,12], directly operating at a biologically meaningful level. As a consequence, representations of tissue structures have shown enhanced explainability [13] and enable the direct interpretation by domain experts. Motivated by these premises, throughout this paper we focus on cell-graphs, assigning cells to nodes and setting edges between adjacent cells.

Generative AI for Biomedicine. In several biomedical settings, further application and development of data-intensive deep learning methods has been hindered mainly by a lack of high-quality annotated samples, as well as by extensive ethical and privacy regulations. Such is the case for DP and the usage of synthetic data has emerged as a promising research direction to address its scarcity of WSIs. Moreover, the pre-processing steps of obtaining entity-graphs from WSIs, such as stain normalization, image segmentation, or entity detection, lack a standardized framework, being cumbersome, time-consuming and handicapping reproducibility [12]. The possibility of directly generating synthetic entity-graphs jointly tackles those limitations.

Several ways to augment medical data have been explored, ranging from adding slightly but naively transformed copies of already existing data [4], which often lead to unexpected distribution shifts, to entire physical simulations [26], which impose a heavy computational burden. Currently, the most promising direction is the use of deep generative models, which seek to generate samples following the same data distribution as their training dataset leveraging on deep learning models. In particular, diffusion models [11,25] emerged as the state-of-the-art approach for deep generative modelling by combining several desirable properties such as training stability, excellent sample quality, easy model scaling, and good distribution coverage. In the specific context of biological structure modelling, these have been applied to tissue slide images [18], outperforming the previously predominant Generative Adversarial Networks [15]. Nevertheless, all the existing approaches have remained at the pixel level, aiming to directly generate new DP images and, thus, suffering of the aforementioned image-based deep learning limitations. By deploying diffusion models at the graph level, we seek to combine the generative potential of the former with the efficient modelling capabilities of relation-aware representations of the latter.

Recently, several graph-based diffusion schemes have been proposed through the adaptation of either score-based generative modelling [14,20] or of discrete denoising diffusion probabilistic models (D3PMs) [2] adapted to graphs [8,27]. For the first class of methods, graphs are embedded in a continuous space and Gaussian noise is used to corrupt its node features and adjacency matrix. This approach neither preserves the inherent sparsity of graphs nor scales for larger instances. Conversely, DiGress, a method of the second class, performs diffusion on fully discrete data structures [27]. Moreover, it also promotes the sparsity of the generated graphs by choosing an adequate noise model, and enhances the expressivity of the used denoising graph neural network (GNN) by adding extra features to its input, leading to state-of-the-art results. For these reasons, DiGress is the graph-based diffusion method that we adopt in this paper.

3   Diffusion Based Cell-Graph Generation of Tertiary Lymphoid Structures

In this section, we start by framing the problem at hand in more detail and then introduce the graph diffusion method adopted in this work.

Problem Formulation. We build a cell-graph by assigning cells as nodes and by setting edges between neighboring cells to represent a proxy for spatial cell-cell interactions, which are solely local. Node features represent cell types that we assume to biologically characterize a cell both anatomically and physiologically. More formally, provided a graph G, we denote by x_i the node attribute (cell type) of node i and by $\mathbf{x}_i \in \{0,1\}^b$ its one-hot encoding, as we consider a total of b different cell types. These are stored in a matrix $\mathbf{X} \in \{0,1\}^{n \times b}$. Similarly, we denote by e_{ij} the edge attribute between the i-th and j-th nodes and by $\mathbf{e}_{ij}$ its one-hot encoding, that is stored in $\mathbf{E}$. As the generative model treats the absence of an edge as a specific edge type, we have $\mathbf{e}_{ij} \in \{0,1\}^{c+1}$, where c is the number of edge types considered and, consequently, $\mathbf{E} \in \{0,1\}^{n \times n \times (c+1)}$. Thus, we unequivocally consider cell-graphs in the form of an unweighted graph $G = (\mathbf{X}, \mathbf{E})$.

In these cell-graphs, we are mostly interested in informative biological structures, such as the TLSs [6,10,16,19,23]. These are composed of clusters of B-cells supported by T-cell compartments and have been shown to be meaningful for medical prognosis in cancer. Their presence in cell-graphs can be measured through the TLS-like organization metric, $\kappa(a)$, allowing the computation of the TLS content given a cell-graph [23]. This computation only considers the edges between B- and T-cells: edges between two cells of the same type are considered a type α edge, while edges between B- and T-cells are considered a type γ_k edge, where k is the number of B-cells adjacent to the vertex B-cell. The classification of different edges of a cell-graph into these classes is illustrated in Fig. 1. Provided a cell-graph, its $\kappa(a)$ value is defined as the proportion of its γ edges whose index is larger than a and, consequently, monotonically decreasing with a:

$$\kappa(a) = \frac{|E| - |E_\alpha| - \sum_{k=0}^{a} |E_{\gamma_k}|}{|E| - |E_\alpha|}, \tag{1}$$

where $|E|$, $|E_\alpha|$, and $|E_{\gamma_k}|$ respectively denote the number of edges, of α edges and of γ_k edges in the given graph [23].

Here, we consider $a \in \{0, \dots, 5\}$, since we verified empirically that $\kappa(a)$ for larger a is non-informative (mostly 0 or, in the rare exceptions, extremely close to that value; check Appendix B for further details). Consequently, each cell-graph is characterized by its *TLS embedding*, $[\kappa(0), \dots, \kappa(5)] \in \mathbb{R}^6$. Using the TLS embedding of cell-graphs, we are able to characterize distributions of such data structures. Based on insights from the paper proposing the TLS embedding [23] and from empirical validation, we can identify two distinct populations of cell-graphs: high TLS content cell-graphs, characterized by $\kappa(2) > 0.05$ and, conversely, low TLS content cell-graphs, which verify $\kappa(1) < 0.05$.

The problem that we consider here is the generation of synthetic data that would be representative of each of the two populations of cell-graphs. Namely, we are interested in building a generative model that can be trained on a few actual samples that are representative of low and high TLS content cell-graphs, respectively, and that can eventually generate new cell-graphs following the same distribution of those two populations.

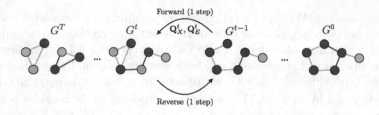

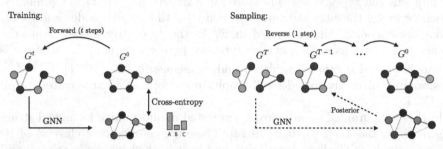

Fig. 2. DiGress is composed of a forward process, whose noise model is modulated by $\mathbf{Q}_E^t$ and $\mathbf{Q}_X^t$, and a reverse process. In the training process, a GNN is trained to predict the clean graph, G^0, that originated its noisy input, G^t. To sample from DiGress, we perform iteratively T reverse steps starting from a fully noisy graph, G^T. In each reverse step, the noisy graph is passed as input to the GNN, whose output is then noised back using the posterior term of the diffusion model. Figure adapted from [27].

Diffusion Based Cell-Graph Generative Model. In our setting, we extend DiGress [27], a D3PM adapted for graphs whose nodes and edges have categorical features, to the TLS cell-graph generation setting. This method is a diffusion model composed of two different processes: *forward* and *reverse*.

In the *forward* process of the diffusion model, we iteratively alter clean graphs, G^0, with noise, until we arrive to its fully corrupted form, G^T, obtaining a trajectory $\{G^0, \ldots, G^T\}$. The noise model adopted by DiGress is independently employed to nodes and edges through transition matrices. Thus, for each node and for each edge, we apply:

$$[\mathbf{Q}_X^t]_{ij} = q(x^t = j | x^{t-1} = i) \quad \text{and} \quad [\mathbf{Q}_E^t]_{ij} = q(e^t = j | e^{t-1} = i), \qquad (2)$$

respectively. Consequently, the categorical distribution that allows us to go one step forward towards the fully noisy graph side of the trajectory is given by:

$$q(G^t | G^{t-1}) = (\mathbf{X}^{t-1} \mathbf{Q}_X^t, \mathbf{E}^{t-1} \mathbf{Q}_E^t). \qquad (3)$$

The adopted noise model follows the marginal scheme [27], which promotes transitions to node and edge types that are more prevalent in the training set. Thus, the transition matrices are given by:

$$\mathbf{Q}_X^t = \alpha^t \mathbf{I} + \beta^t \mathbf{1}_\mathbf{A} \mathbf{m}_X' \quad \text{and} \quad \mathbf{Q}_E^t = \alpha^t \mathbf{I} + \beta^t \mathbf{1}_\mathbf{B} \mathbf{m}_E', \qquad (4)$$

where α^t transitions from 1 to 0 with t according to the popular cosine scheduling. Then, $\beta^t = 1 - \alpha^t$, $\mathbf{1}_A \in \{1\}^b$, $\mathbf{1}_B \in \{1\}^c$ are filled with ones, and

$\mathbf{m}'_X \in \mathbb{R}^b, \mathbf{m}'_E \in \mathbb{R}^c$ are row vectors ($'$ denotes transposition) filled with the marginal distributions of node and edge types, respectively.

In the *reverse* process of DiGress, a graph transformer network (i.e., a type of GNN) is trained to recursively backtrack the trajectories generated by the forward process. To overcome the limited representational power of GNNs, graph-theoretic (cycles and spectral) auxiliary features are added to the transformer model for each reverse step.

With this setup, DiGress is trained by repeatedly performing the following steps: pick a clean graph from the training set, G^0, and perform t forward steps, where t is chosen randomly between 1 and T, obtaining G^t. Importantly, it is possible to perform the t forward steps in a single closed-form computation, given $\mathbf{Q}_X^1, \ldots, \mathbf{Q}_X^t, \mathbf{Q}_E^1, \ldots, \mathbf{Q}_E^t$. Then, the GNN, which is the only trainable component of DiGress, is fed with G^t and predicts probability distributions over the nodes and edges types for each node and edge of G^0, being then subject to binary cross-entropy loss. After having trained the GNN, we can sample from DiGress. First, we sample a fully noisy graph, G^T, from the limit distribution of the forward process, which is fixed by definition. Then, to obtain a new clean sample, we iteratively denoise G^T by performing T reverse steps. Each of the reverse steps consists of a GNN prediction given the noisy graph as input, followed by a "noise back" step using the posterior term of DiGress. This posterior term can be computed in closed-form provided the diffusion model transition matrices. Both the training and the sampling procedures from DiGress are illustrated in Fig. 2.

Altogether, the diffusion-based graph generative model built on Digress permits to infer the distribution of the data that the model has been trained on. As a result, when properly trained on actual samples of TLS cell-graphs, the generation of novel graphs is simply achieved by sampling from the learned cell-graph distribution.

4   Experiments

In this section, we first describe the data preprocessing steps, followed by an evaluation of the quality of the cell-graphs generated using DiGress, comparing it to a non deep learning baseline. Then, we illustrate the utility of the trained DiGress models for data augmentation in a TLS content classification task. Finally, we report the computational resources required for the previous tasks.

Data and Preprocessing. We create our actual datasets[1] following the classical procedure for cell-graph extraction in digital pathology [7,23]. The first step consists of segmenting the cells of the WSI. We set the detected cells to nodes, whose types form node features. We consider 9 different cell types, identified in Fig. 1. This imposes $b = 9$ in the generative model. To obtain the whole slide graph, edges are set between adjacent nodes using Delaunay triangulation and

---

[1] The used datasets resulted from a collaboration with Centre Hospitalier Universitaire Vaudois and are not publicly available.

edge thresholding (edges longer than 30 μm are ignored). As we only consider one edge type, we have $c = 1$. Finally, from each whole slide graph, we extract several 4-hop subgraphs centered on B-cells. These subgraphs are then assigned to two datasets: $\mathcal{D}_{\text{Real}}^{\sim\text{TLS}}$ composed of cell-graphs with low TLS content ($\kappa(1) < 0.05$) and $\mathcal{D}_{\text{Real}}^{\text{TLS}}$ with high TLS content cell-graphs ($\kappa(2) > 0.05$). $\mathcal{D}_{\text{Real}}^{\sim\text{TLS}}$ and $\mathcal{D}_{\text{Real}}^{\text{TLS}}$ contain 3020 and 5042 cell-graphs, respectively, with varying number of nodes between 20 and 103. We also define $\mathcal{D}_{\text{Real}} = \mathcal{D}_{\text{Real}}^{\sim\text{TLS}} \cup \mathcal{D}_{\text{Real}}^{\text{TLS}}$.

Baseline. We define a non deep learning based but reasonable baseline that captures the 1-hop dependencies of cell-graphs. This model learns the following marginal distributions from the respective training set: number of nodes per graph (distribution *I*), cell types (distribution *II*), and edge types given the cell types of the edge vertices (distribution *III*). Importantly, in the scope of this paper, *I* and *II* are categorical distributions, while *III* consists of a Bernoulli distribution. Thereafter, the sampling procedure of a graph from the trained baseline model is as follows: *i)* sample a number of nodes from *I*; *ii)* for each of those nodes, sample a cell type from *II*; and, finally, *iii)* sample an edge type between every two nodes from *III*. Noteworthy, for a given edge, the vertices cell types are, at that point, already known from step *ii)*.

Evaluation of Synthetic Cell-Graphs Distributions. We separately train a DiGress model and a baseline model for each of the datasets $\mathcal{D}_{\text{Real}}^{\sim\text{TLS}}$ and $\mathcal{D}_{\text{Real}}^{\text{TLS}}$, yielding a total of 4 trained models. We sample from each of the trained models, obtaining $\mathcal{D}_{\text{DiGress}}^{\sim\text{TLS}}$, $\mathcal{D}_{\text{DiGress}}^{\text{TLS}}$, $\mathcal{D}_{\text{Baseline}}^{\sim\text{TLS}}$, and $\mathcal{D}_{\text{Baseline}}^{\text{TLS}}$, each with 5000 cell-graphs. Then, we compare these generated datasets with the respective training sets through their distributions, in a biologically meaningful way: using the TLS embedding distributions. This comparison is made through four different metrics typically adopted in synthetic data evaluation [22], as they emphasize different similarities between probability distributions. In particular, the Kolmogorov-Smirnov test (KS) focuses on similarity of distributions based on the maximum vertical distance between their cumulative distribution functions. The Wasserstein distance (WD) captures differences in the shape, location, and spread of distributions by measuring the minimum amount of work required for transformation of one into the other. The Jensen-Shannon divergence (D_{JS}) captures dissimilarity based on information content by comparing the divergence of each distribution from their average distribution. Lastly, the maximum mean discrepancy (MMD) evaluates the difference in means of data representations transformed using a kernel function. The results, presented in Table 1, show that the intricate biological dependencies found in a cell-graph go beyond 1-hop relations, since those are the ones that the baseline is able to capture accurately. In contrast, by capturing higher order and more complex dependencies, DiGress significantly outperforms the baseline in both training sets. This fact highlights the pertinence of graph-based diffusion models to capture cell-graph distributions.

Table 1. Evaluation of the generated datasets through TLS embedding distribution comparison with the real TLS dataset, $\mathcal{D}_{\text{Real}}^{\text{TLS}}$, (*left*) and ~TLS dataset, $\mathcal{D}_{\text{Real}}^{\sim\text{TLS}}$, (*right*). The TLS embedding entries are denoted by $\kappa(a)$. KS stands for Kolmogorov-Smirnov test, WD for Wasserstein distance, D_{JS} for Jensen-Shannon divergence, and MMD for maximum mean discrepancy.

		TLS						~TLS					
		$\kappa(0)$	$\kappa(1)$	$\kappa(2)$	$\kappa(3)$	$\kappa(4)$	$\kappa(5)$	$\kappa(0)$	$\kappa(1)$	$\kappa(2)$	$\kappa(3)$	$\kappa(4)$	$\kappa(5)$
KS	Baseline	0.127	0.092	**0.088**	0.105	0.114	0.080	0.136	0.398	0.104	0.018	0.002	**0.000**
	DiGress	**0.060**	**0.062**	0.110	**0.051**	**0.049**	**0.057**	**0.097**	**0.160**	**0.007**	**0.000**	**0.000**	0.000
WD	Baseline	0.028	**0.041**	**0.044**	0.034	0.016	0.009	0.064	0.079	0.013	0.002	0.000	**0.000**
	DiGress	**0.019**	0.044	0.045	**0.024**	**0.016**	**0.006**	**0.053**	**0.017**	**0.001**	**0.000**	**0.000**	0.000
D_{JS}	Baseline	0.199	0.202	0.228	0.144	0.120	0.084	0.168	0.419	0.190	0.079	0.028	**0.000**
	DiGress	**0.087**	**0.126**	**0.215**	**0.093**	**0.088**	**0.071**	**0.136**	**0.265**	**0.053**	**0.000**	**0.000**	0.000
MMD	Baseline	0.062	0.076	0.071	0.068	0.049	0.024	0.095	0.602	0.059	0.002	0.000	**0.000**
	DiGress	**0.012**	**0.024**	**0.033**	**0.013**	**0.012**	**0.013**	**0.051**	**0.086**	0.000	0.000	0.000	0.000

Synthetic Cell-Graphs in Data Augmentation. To illustrate the utility of generative models in downstream tasks, we explore data augmentation with synthetic cell-graphs in a scarcity regime of real data, a setting often found in biomedical applications. In particular, we consider a binary classification task where a graph convolutional network (GCN) is trained to predict if a cell-graph has a high or low TLS content. We build two datasets with freshly collected cell-graphs from the same tissue slide cell-graphs, $\mathcal{D}_{\text{Real}}'^{\sim\text{TLS}}$ and $\mathcal{D}_{\text{Real}}'^{\text{TLS}}$, both with 100 cell-graphs. We split the dataset $\mathcal{D}_{\text{Real}}' = \mathcal{D}_{\text{Real}}'^{\sim\text{TLS}} \cup \mathcal{D}_{\text{Real}}'^{\text{TLS}}$ into balanced and equally sized training and test sets. Finally, during the training of the GCN, we augment the training set with positive and negative cell-graphs (by the same amount) coming from three different origins: *i)* real data, $\mathcal{D}_{\text{Real}}^{\sim\text{TLS}}$ and $\mathcal{D}_{\text{Real}}^{\text{TLS}}$; *ii)* synthetic cell-graphs generated by DiGress, $\mathcal{D}_{\text{DiGress}}^{\sim\text{TLS}}$, and $\mathcal{D}_{\text{DiGress}}^{\text{TLS}}$; and *iii)* synthetic cell-graphs generated by the baseline model, $\mathcal{D}_{\text{Baseline}}^{\sim\text{TLS}}$, and $\mathcal{D}_{\text{Baseline}}^{\text{TLS}}$. The magnitude of data augmentation ranges from $1\times$ (which means no augmentation) to $40\times$ (which means that the augmented training set is 40 times larger than the original training set).

Since the adopted criteria to consider high and low TLS content cell-graphs consists of a 2-hop condition, we consider a 2-layer GCN for the illustrative binary graph classification task. This model is trained using binary cross-entropy until a maximum of 5000 epochs, or less in case the cross-entropy of the model does not increase in the validation set for 100 epochs (early-stopping). For each type and magnitude of augmentation, we tune the hyperparameters within a grid: learning rate takes the values 0.01, 0.001, or 0.0001; the dimension of the graph embeddings in the hidden layers is 8 or 16; the dropout rate is 0.2 or 0.5. We find the best hyperparameter combination considering the mean AUROC of the trained models in the corresponding validation sets in a 5-fold stratified cross-validation procedure. The 5 models that share the best hyperparameter configuration are then evaluated at the test set. The obtained results can be found in Table 2.

Table 2. Validation (top) and test (bottom) AUROC (mean ± standard error of the mean) obtained for the 5 models sharing the best hyperparameter combination for each type and magnitude of data augmentation.

Val.	1×	2×	3×	5×	10×	20×	40×
Real	**0.970** ± 0.010	**0.976** ± 0.008	**0.982** ± 0.009	**0.990** ± 0.006	**0.990** ± 0.005	**0.996** ± 0.004	**0.994** ± 0.004
DiGress	**0.970** ± 0.010	0.972 ± 0.015	0.978 ± 0.013	0.982 ± 0.011	0.974 ± 0.011	0.976 ± 0.017	0.978 ± 0.015
Baseline	**0.970** ± 0.010	0.960 ± 0.019	0.944 ± 0.028	0.950 ± 0.024	0.932 ± 0.032	0.934 ± 0.031	0.938 ± 0.023
Test	1×	2×	3×	5×	10×	20×	40×
Real	**0.921** ± 0.004	0.926 ± 0.005	**0.952** ± 0.006	**0.947** ± 0.003	**0.956** ± 0.010	**0.962** ± 0.008	**0.966** ± 0.010
DiGress	**0.921** ± 0.004	**0.946** ± 0.007	0.943 ± 0.010	**0.947** ± 0.010	0.945 ± 0.010	0.931 ± 0.009	0.939 ± 0.009
Baseline	**0.921** ± 0.004	0.938 ± 0.009	0.927 ± 0.007	0.932 ± 0.005	0.928 ± 0.008	0.927 ± 0.007	0.922 ± 0.004

As a general observation, data augmentation leads to improvements in model performance, which is a consequence of the regularization effect of data augmentation in learning tasks. However, this improvement differs with the origin of the added cell-graphs. For real data, we observe that overall the stronger the augmentation magnitude, the better the performance. This result is an expected consequence of the augmentation cell-graphs following exactly the same distribution as the ones found in the training and test set: despite being different, all of them were extracted from the same tissue slides. When the augmentation is performed with DiGress generated cell-graphs, the model test performance increases as well, but not as much as by using real data. This result comes as a consequence of the samples generated by DiGress not following exactly the same distribution of $\mathcal{D}_{\text{Real}}$ (or, equivalently, $\mathcal{D}'_{\text{Real}}$), as shown in Table 1. By the same token, the performance improvement brought by the augmentation with the baseline model cell-graphs is more subtle. We can finally note that, for the augmentations with synthetically generated cell-graphs, there is no monotonic trend of performance improvement with the magnitude of data augmentation. We attribute the test AUROC deterioration for augmentations larger than 5× for DiGress and 2× for the baseline model to an excessive exposition of the model to synthetic data, which unavoidably leads to distribution shifts in the total training set. The performance deterioration is detected at smaller augmentation magnitudes for the baseline model as it causes larger distribution shifts.

Computational Resources. The training of the deep learning models was carried out using a single Tesla V100 GPU. The training of the DiGress models on the $\mathcal{D}_{\text{Real}}^{\sim\text{TLS}}$ and $\mathcal{D}_{\text{Real}}^{\text{TLS}}$ datasets required 19 and 30 h, respectively. The generation of

$\mathcal{D}_{\text{DiGress}}^{\sim \text{TLS}}$ and $\mathcal{D}_{\text{DiGress}}^{\text{TLS}}$ consumed 17 h each, approximately. This sampling procedure was performed in the same single GPU. For the baseline model, its training on the $\mathcal{D}_{\text{Real}}^{\sim \text{TLS}}$ and $\mathcal{D}_{\text{Real}}^{\text{TLS}}$ datasets only took 14 and 20 min, respectively. The generation of $\mathcal{D}_{\text{Baseline}}^{\sim \text{TLS}}$ and $\mathcal{D}_{\text{Baseline}}^{\text{TLS}}$ requires only 20 s for each of them. Finally, for the training of the GCN used in the classification task, 1260 runs were carried out to cover all the hyperparameter configurations explored, cross-validation folds, and sources of augmentation. The longest of them took approximately 17 min.

5   Conclusion

This paper provides, to the best of our knowledge, the first application of graph-based diffusion models to the digital pathology domain. We leverage on a current state-of-the-art diffusion model for graphs with categorical node and edges features, DiGress, to successfully generate high and low TLS content cell-graphs. Our results highlight the relevance of capturing higher order relations to accurately model the complex biological dependencies that rule the digital pathology data. Furthermore, we demonstrate the practical value of generative models as a means of augmenting this data type. As future work, we are looking at properly incorporating *prior* biological knowledge to the generative model in order to further enhance the promising results obtained in this work. The main limitation of the application of current state-of-the-art graph-based generative models in whole-slide pathology images is related with their limited scalability: developing implementations that can generate large graphs is an important future direction.

Acknowledgements. The authors would like to thank Alexandre Wicky, Michel A. Cuendet and Olivier Michielin for the fruitful discussions about the biological interpretation of cell-graph representations and to Beril Besbinar for the help in data preprocessing and useful methodological suggestions.

A   Examples of High and Low TLS Content Subgraph

In this section we present examples of high and low TLS content cell-graphs to provide further intuition on the difference between the two classes. It is possible to observe a distinct cluster of B-cells surrounded by supporting T-cells for high TLS content cell-graphs, whereas the low TLS content counterparts have a more heterogeneous composition and configuration, as illustrated in Fig. 3.

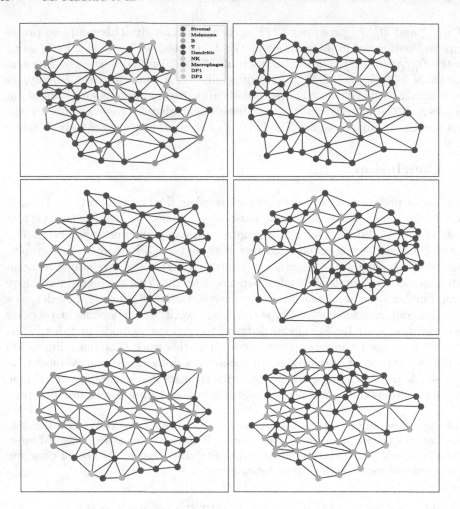

Fig. 3. Subgraphs extracted from whole-slide cell-graphs. *Left:* Low TLS content cell-graphs. *Right:* High TLS content cell-graphs. These cell-graphs are similar to the one found at the center of Fig. 1.

B   Distributions of TLS Embeddings

To further illustrate the results obtained in Sect. 4, we provide the TLS embeddings distributions for the different datasets considered. The distributions obtained for the large real dataset, for the dataset sampled from DiGress, and for the dataset sampled from the baseline model are depicted in Fig. 4, Fig. 5, and Fig. 6, respectively.

In Fig. 4, it is possible to observe the sharp transitions imposed by the low TLS content criterion ($\kappa(1) < 0.05$) for the $\mathcal{D}_{\text{Real}}^{\sim\text{TLS}}$ dataset and by the high TLS content criterion ($\kappa(2) > 0.05$) for the $\mathcal{D}_{\text{Real}}^{\text{TLS}}$. This transition is not well captured by the generative models, explaining their worse performance for $\kappa(1)$ (in $\sim$TLS) and $\kappa(2)$ (in TLS), namely for DiGress (see Table 1).

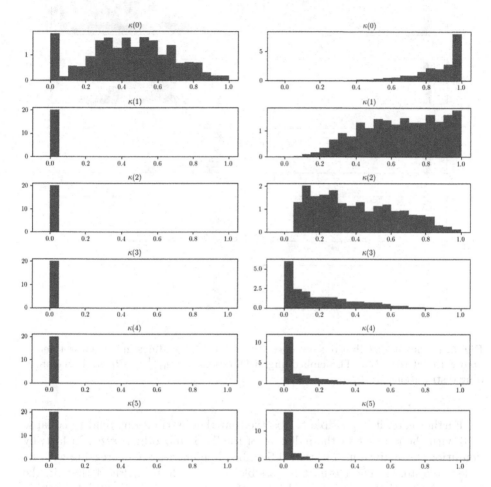

Fig. 4. Empirical distributions obtained for the TLS embeddings in the real data. *Left:* TLS embedding distributions for $\mathcal{D}_{\text{Real}}^{\sim\text{TLS}}$. *Right:* TLS embedding distributions for $\mathcal{D}_{\text{Real}}^{\text{TLS}}$.

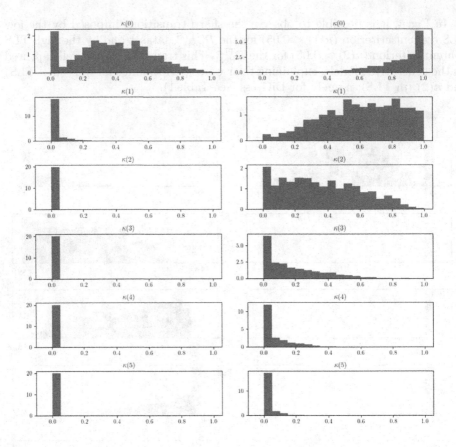

Fig. 5. Empirical distributions obtained for the TLS embeddings in the dataset generated with DiGress. *Left:* TLS embedding distributions for $\mathcal{D}_{\text{DiGress}}^{\sim\text{TLS}}$. *Right:* TLS embedding distributions for $\mathcal{D}_{\text{DiGress}}^{\text{TLS}}$.

Furthermore, it is possible to observe that the distributions tend to collapse to 0 with the increase of the index, a, of the TLS embedding, $\kappa(a)$. In fact, by definition, for a given graph, the TLS embedding entries are decreasing with its index (see Eq. 1). Therefore, it is possible to conclude from Fig. 4 that, in the vast majority of the cases, no relevant information remains in $\kappa(a)$ for $a > 5$, supporting our decision of discarding their computation.

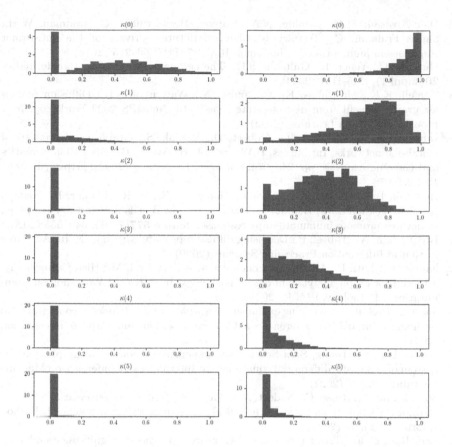

Fig. 6. Empirical distributions obtained for the TLS embeddings in the dataset generated with the baseline model. *Left:* TLS embedding distributions for $\mathcal{D}_{\text{Baseline}}^{\sim\text{TLS}}$. *Right:* TLS embedding distributions for $\mathcal{D}_{\text{Baseline}}^{\text{TLS}}$.

References

1. Ahmedt-Aristizabal, D., Armin, M.A., Denman, S., Fookes, C., Petersson, L.: A survey on graph-based deep learning for computational histopathology. Comput. Med. Imaging Graph. **95**, 102027 (2022)
2. Austin, J., Johnson, D.D., Ho, J., Tarlow, D., van den Berg, R.: Structured denoising diffusion models in discrete state-spaces. In: Advances in Neural Information Processing Systems (2021)
3. Bera, K., Schalper, K.A., Rimm, D.L., Velcheti, V., Madabhushi, A.: Artificial intelligence in digital pathology–new tools for diagnosis and precision oncology. Nat. Rev. Clin. Oncol. **16**(11), 703–715 (2019)
4. Chlap, P., Min, H., Vandenberg, N., Dowling, J., Holloway, L., Haworth, A.: A review of medical image data augmentation techniques for deep learning applications. J. Med. Imaging Radiat. Oncol. **65**(5), 545–563 (2021)
5. De Jong, H.: Modeling and simulation of genetic regulatory systems: a literature review. J. Comput. Biol. **9**(1), 67–103 (2002)

6. Dieu-Nosjean, M.C., Giraldo, N.A., Kaplon, H., Germain, C., Fridman, W.H., Sautès-Fridman, C.: Tertiary lymphoid structures, drivers of the anti-tumor responses in human cancers. Immunol. Rev. **271**(1), 260–275 (2016)
7. Gunduz, C., Yener, B., Gultekin, S.H.: The cell graphs of cancer. Bioinformatics **20**(suppl-1), i145–i151 (2004)
8. Haefeli, K.K., Martinkus, K., Perraudin, N., Wattenhofer, R.: Diffusion models for graphs benefit from discrete state spaces. In: NeurIPS 2022 Workshop: New Frontiers in Graph Learning (2022)
9. van Helden, J., Wernisch, L., Gilbert, D., Wodak, S.: Graph-based analysis of metabolic networks. In: Mewes, H.W., Seidel, H., Weiss, B. (eds.) Bioinformatics and Genome Analysis, pp. 245–274. Springer, Heidelberg (2002). https://doi.org/10.1007/978-3-662-04747-7_12
10. Helmink, B.A., Reddy, S.M., Gao, J., Zhang, S., Basar, R., Thakur, R., Yizhak, K., Sade-Feldman, M., Blando, J., Han, G., et al.: B cells and tertiary lymphoid structures promote immunotherapy response. Nature **577**(7791), 549–555 (2020)
11. Ho, J., Jain, A., Abbeel, P.: Denoising diffusion probabilistic models. In: Advances in Neural Information Processing Systems (2020)
12. Jaume, G., Pati, P., Anklin, V., Foncubierta, A., Gabrani, M.: HistoCartography: a toolkit for graph analytics in digital pathology. In: MICCAI Workshop on Computational Pathology. PMLR (2021)
13. Jaume, G., et al.: Quantifying explainers of graph neural networks in computational pathology. In: IEEE Conference on Computer Vision and Pattern Recognition (2021)
14. Jo, J., Lee, S., Hwang, S.J.: Score-based generative modeling of graphs via the system of stochastic differential equations. In: International Conference on Machine Learning. PMLR (2022)
15. Jose, L., Liu, S., Russo, C., Nadort, A., Di Ieva, A.: Generative adversarial networks in digital pathology and histopathological image processing: a review. J. Pathol. Inform. **12**(1), 43 (2021)
16. Lee, H.J., et al.: Tertiary lymphoid structures: prognostic significance and relationship with Tumour-infiltrating lymphocytes in triple-negative breast cancer. J. Clin. Pathol. **69**(5), 422–430 (2016)
17. Li, M.M., Huang, K., Zitnik, M.: Graph representation learning in biomedicine and healthcare. Nature Biomed. Eng. **6**(12), 1353–1369 (2022)
18. Moghadam, P.A., et al.: A morphology focused diffusion probabilistic model for synthesis of histopathology images. In: Proceedings of the IEEE/CVF Winter Conference on Applications of Computer Vision (2023)
19. Munoz-Erazo, L., Rhodes, J.L., Marion, V.C., Kemp, R.A.: Tertiary lymphoid structures in cancer-considerations for patient prognosis. Cell. Mol. Immunol. **17**(6), 570–575 (2020)
20. Niu, C., Song, Y., Song, J., Zhao, S., Grover, A., Ermon, S.: Permutation invariant graph generation via score-based generative modeling. In: International Conference on Artificial Intelligence and Statistics. PMLR (2020)
21. Pitzalis, C., Jones, G.W., Bombardieri, M., Jones, S.A.: Ectopic lymphoid-like structures in infection, cancer and autoimmunity. Nat. Rev. Immunol. **14**(7), 447–462 (2014)
22. Qian, Z., Cebere, B.C., van der Schaar, M.: Synthcity: facilitating innovative use cases of synthetic data in different data modalities. arXiv preprint arXiv:2301.07573 (2023)
23. Schaadt, N.S., et al.: Graph-based description of tertiary lymphoid organs at single-cell level. PLoS Comput. Biol. **16**(2), e1007385 (2020)

24. Serag, A., et al.: Translational AI and deep learning in diagnostic pathology. Front. Med. **6**, 185 (2019)
25. Sohl-Dickstein, J., Weiss, E., Maheswaranathan, N., Ganguli, S.: Deep unsupervised learning using nonequilibrium thermodynamics. In: International Conference on Machine Learning. PMLR (2015)
26. Tang, C., Vishwakarma, S., Li, W., Adve, R., Julier, S., Chetty, K.: Augmenting experimental data with simulations to improve activity classification in healthcare monitoring. In: 2021 IEEE radar conference (RadarConf21). IEEE (2021)
27. Vignac, C., Krawczuk, I., Siraudin, A., Wang, B., Cevher, V., Frossard, P.: DiGress: discrete Denoising diffusion for graph generation. In: International Conference on Learning Representations (2022)
28. Wu, Z., et al.: Graph deep learning for the characterization of tumour microenvironments from spatial protein profiles in tissue specimens. Nature Biomed. Eng. **6**, 1435–1448 (2022)

Prior-RadGraphFormer:
A Prior-Knowledge-Enhanced
Transformer for Generating Radiology
Graphs from X-Rays

Yiheng Xiong[1]([✉]), Jingsong Liu[1], Kamilia Zaripova[1], Sahand Sharifzadeh[2], Matthias Keicher[1], and Nassir Navab[1]

[1] Computer Aided Medical Procedures, Technische Universität München, Munich, Germany
yiheng.xiong@tum.de
[2] Ludwig Maximilians Universität München, Munich, Germany

Abstract. The extraction of structured clinical information from free-text radiology reports in the form of radiology graphs has been demonstrated to be a valuable approach for evaluating the clinical correctness of report-generation methods. However, the direct generation of radiology graphs from chest X-ray (CXR) images has not been attempted. To address this gap, we propose a novel approach called Prior-RadGraphFormer that utilizes a transformer model with prior knowledge in the form of a probabilistic knowledge graph (PKG) to generate radiology graphs directly from CXR images. The PKG models the statistical relationship between radiology entities, including anatomical structures and medical observations. This additional contextual information enhances the accuracy of entity and relation extraction. The generated radiology graphs can be applied to various downstream tasks, such as free-text or structured reports generation and multi-label classification of pathologies. Our approach represents a promising method for generating radiology graphs directly from CXR images, and has significant potential for improving medical image analysis and clinical decision-making. Our code is open sourced at https://github.com/xiongyiheng/Prior-RadGraphFormer.

Keywords: Radiology Graph Generation · Transformer · Prior Knowledge

---

Y. Xiong, J. Liu and K. Zaripova—These authors contribute equally to this work and share first authorship.
M. Keicher and N. Navab—These authors share last authorship.

Supplementary Information The online version contains supplementary material available at https://doi.org/10.1007/978-3-031-55088-1_5.

1   Introduction

In recent years, deep learning (DL) methods have significantly improved the computer-assisted diagnosis of chest X-ray (CXR) images. DL methods have been used in multiple ways. For instance, Ma and Lv [19] used a Swin transformer with fully-connected layers to classify CXR images as either normal or indicative of pneumonia. Cicero et al. [3] modeled the diagnosis process as a multi-label classification problem and used GoogleNet to classify CXR images into six categories. In comparison to these classification methods, generated free-text radiology reports can provide more comprehensive information about the impression and findings [9]. Shin et al. [23] pioneered CNN-RNN for automatic text generation. Wang et al. [25] introduced TieNet, generating reports and detecting thorax diseases. Hou et al. [6] proposed RATCHET, using transformer and attention to generate reports from CXR images. Kaur et al. [12] used contextual word representations for useful radiological reports. Cao et al. [1] proposed Multi-modal Memory Transformer Network for consistent medical reports. Existing approaches for automating medical reporting rely on generating free text, posing challenges for clinical evaluation [13,20]. To tackle this, ImaGenome [27] and RadGraph Benchmark [9] were introduced to extract structured clinical information from free-text radiology reports, represented as a radiology graph. Each node in a radiology graph corresponds a unique entity, such as an object with bounding boxes annotations or an attribute in ImaGenome or, an anatomical structure or an observation and its presence and uncertainty in RadGraph Benchmark. Although these graph representations of reports have been used to evaluate the clinical correctness of reports [28], generating radiology graphs directly from CXR images has not been explored. In contrast, diverse interactions between object pairs in natural images in the form of scene graph generation have been extensively investigated [17]. Li et al. [14], and Lu et al. [18] employed two-stage methods to propose dense relationships between predicted connected object pairs. In recent work, Shit et al. [24] introduced Relationformer, a unified one-stage framework based on DETR [2] that facilitates the end-to-end generation of graphs from images. It is a state-of-the-art method for detecting and generating graphs from natural images. However, it requires bounding boxes for the detected objects (nodes of the graph) and is not directly applicable to radiology graphs since some nodes (entities) in radiology graphs do not have exact locations, such as "left" or "clear", and most datasets do not provide bounding boxes annotations.

Therefore, we propose a detection-free method, Prior-RadGraphFormer, to generate radiology graphs directly from CXR images without requiring bounding boxes for each entity. The method incorporates prior knowledge in the form of probabilistic knowledge graphs (PKG) [22] that model the statistical relationship between anatomies and pathological observations. Experimental results show that Prior-RadGraphFormer achieves competitive results in the radiology-graph-generation task. Moreover, the generated graphs can be used for multiple downstream tasks such as generating free-text reports based on predefined rules, cheXpert labels [8] classification, and populating templates for structured reporting.

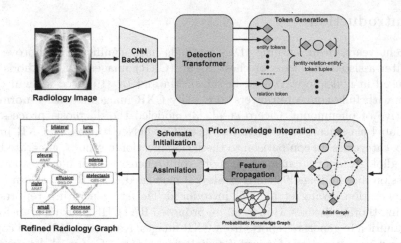

Fig. 1. Prior-RadGraphFormer architecture. A CNN backbone extracts radiology image features, which are fed into the detection transformer (DETR) [2] to generate multiple entity tokens and one relation token. Then an initial graph is constructed using entity and relation tokens. Each node is represented by a valid entity token and each edge is represented by a corresponding [entity-relation-entity]-token tuple. The initial graph is incorporated with prior knowledge and refined to output the final radiology graph.

In summary, our contributions are the following: 1) proposing a novel detection-free method that generates radiology graphs directly from CXR images; 2) enhancing this method by incorporating prior knowledge, leading to improved performance; 3) extensively evaluating our method using RadGraph metrics and the two downstream tasks of report generation and multi-label classification of pathologies.

2   Method

We define radiology graph generation as the process of transforming CXR images into a graph structure introduced by RadGraph Benchmark [9] that represents the content of a radiology report describing the image. Each node in the graph corresponds to a unique entity, such as an anatomical structure or an observation, and its presence and uncertainty. The edges between nodes indicate relationships between these entities. A visual depiction of the proposed method is shown in Fig. 1. Prior-RadGraphFormer consists of two main components: Rad-GraphFormer and Prior Knowledge Integration.

2.1   RadGraphFormer

Shit et al. [24] proposed Relationformer, an image-to-graph framework that leverages direct set-based object prediction and incorporates the interaction among the objects to learn an object-relation representation jointly. Given an input image, Relationformer initially outputs a set of discrete object tokens ([obj]-tokens) and a relation token ([rln]-token) where the token embeddings come from applying cross attention between a set of pre-defined and randomly initialized token vectors to input image patches. Relationformer then predicts the corresponding class label and the location of the bounding boxes for each object. In addition, Relationformer predicts a relation label for each pair of detected objects by concatenating each pair of [obj]-tokens with the [rln]-token and applying a relation prediction head on top.

In our model, we adapt Relationformer as the radiology graph generation backbone. We modify the entity prediction module by replacing the bounding box prediction head of it with the uncertainty prediction head due to the fact that radiology graphs do not contain bounding boxes but uncertainty information for each entity. Specifically, the entity prediction module comprises two distinct components. The first component is responsible for entity classification. The second component is responsible for predicting the uncertainty associated with each entity. Both are presented by one linear layer. To reflect these changes in terminology, we adjust the naming convention of the model's tokens, from [obj]-token to [ent]-token. Moreover, instead of using ResNet50 as the CNN backbone, we utilize DenseNet121 [7] based on its widespread adoption and effectiveness for processing CXR images [6,13].

2.2   Prior Knowledge Integration

Vanilla RadGraphFormer uses the concatenation of a pair of [ent]-tokens and a shared [rln]-token ($\{ent_i, r, ent_j\}_{i \neq j}$), followed by an MLP, for relation prediction. Although the transformer-based model inherently considers the context of the tokens, we argue that these representations may not adequately capture the complexity of the relationships. Consequently, this limited representation may lead to an incomplete or inaccurate understanding of the context, resulting in sub-optimal relation and/or entity prediction.

Following [22], in order to address this issue, we propagate higher-level prior knowledge in RadGraphFormer, as shown in the lower part of Fig. 1. To this end, we first construct an initial embedding graph G from all the valid ent_is as nodes. Each edge in the graph is represented by $edg_{ij} = MLP_{proj}(\{ent_i, r, ent_j\}_{i \neq j})$, where MLP_{proj} helps project the concatenated edge features to the same dimension of node features. This results in a fully-connected bi-directional graph. Then a stack of graph transformers GTs [29] are utilized to propagate features of both nodes and edges in G. In order to allow the storage and propagation of relational knowledge within our framework, we create a randomly initialized representation for each class as a trainable parameter s_c called schemata [22] that interacts with the outputs of the graph transformer. We then apply multiple assimilation steps [22]. The assimilation step is a process where the outputs of GT attend to s_cs such that the attention coefficients predict the classification outputs for each node/edge from the GT and the attention values are propagated to the output of GT. It is important to note that the attention coefficients are supervised by the ground truth labels for each entity/relation during the training. Details can be found in the appendix.

2.3   Training and Inference

During training, we apply two supervisions for entity classes. One is for [ent]-tokens generated from DETR and the other is for the attention coefficients of nodes during each assimilation step. The latter one can be regarded as an additional entity class supervision. Entity uncertainty is supervised via [ent]-tokens and relation is supervised via the attention coefficients of edges during each assimilation step. During inference, we evaluate entity metrics solely based on [ent]-tokens. As for relation inference, we use the attention coefficients of edges from the last assimilation step as the classification output.

3   Experiments

3.1   Datasets

Our dataset comprises MIMIC-CXR-JPG v2.0.0 [4,10,11], which contains both imaging studies and free-text reports, and RadGraph Benchmark [9]. Our training set includes around 220,000 ground truth graph annotations obtained by RadGraph Benchmark, paired with the corresponding frontal CXR images. The validation set consists of 500 ground truth graph annotations obtained by board-certified radiologists and the corresponding frontal CXR images. There are 229 classes of entities after mapping, three levels of uncertainty, and three classes of relation. Details can be found in the appendix.

3.2   Implementation Details

All networks including the backbone are trained from scratch with PyTorch 1.12.0 and CUDA 11.6 on a single NVIDIA A40 until convergence. A batch size of 32 is chosen, and AdamW [16] optimizer with a learning rate of 1e-4 is utilized. To address the issue of imbalanced labels in entity class prediction, we employ focal loss [15], while cross entropy loss [5] is used for entity uncertainty prediction. For relation prediction, we utilize stochastic relation loss [24], with a foreground-to-background edge ratio of 1:3. The weights between entity classification loss, uncertainty loss, and relation loss are 1, 1, and 3. Additionally, two assimilation steps are performed during the prior knowledge integration process per iteration. CXR images are preprocessed following [13].

3.3   Evaluation

We evaluate our model similarly to RadGraph Benchmark [9] considering micro F1-score. A predicted entity is regarded as a true positive if its predicted entity class and uncertainty level are correct. For relation evaluation, a predicted relation is considered a true positive if its head entity, tail entity, and relation type are predicted correctly. To show the use cases for radiology graph generation in this work, we perform the evaluation for the two downstream tasks: free-text report generation and cheXpert labels [8] classification. Specifically, we generate free-text clinical reports from generated radiology graphs based on pre-defined rules.

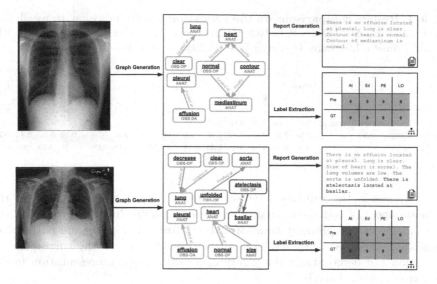

Fig. 2. Qualitative results. On the left and middle are the CXR images and the corresponding predicted radiology graphs by our model. On the right, we present the outputs of two downstream tasks derived via the predicted graph: free-text clinical report and pathologies classification result. True positives, false negatives, and false positives are represented by green, gray, and red, respectively. (Color figure online)

4   Results and Discussion

4.1   Main Results

Given that no prior work has addressed our task, we create a baseline that can be evaluated via RadGraph metrics. The baseline is constructed by two separate pretrained models, where RATCHET [6] is used to predict free-text reports from CXR images and followed by RadGraph Benchmark [9] to produce radiology graphs from predicted reports. Note that radiology graphs generated from ground truth reports via pretrained RadGraph Benchmark are essentially ground truth labels as discussed in Sect. 3.1 hence they are not directly comparable with our results. Additionally, we compare the performance of vanilla RadGraph-Former with Prior-RadGraphFormer. As is shown in Table 1, both vanilla Rad-GraphFormer and Prior-RadGraphFormer demonstrate superior performance over the baseline in all RadGraph metrics. In particular, Prior-RadGraphFormer shows better performance than vanilla RadGraphFormer, indicating the significance of leveraging prior knowledge for radiology graphs generation.

4.2   Downstream Tasks

For downstream tasks, we compare our methods with RATCHET [6] using free-text NLP metrics and cheXpert label classification F1 score. However, please note that there may be variations in the numerical results of RATCHET due

Table 1. Results comparisons on the validation set. (P: precision, R: recall; B-1: BLEU-1, MET.: METEOR, R_L: ROUGE_L scores; At: Atelectasis, Ed: Edema, PE: Pleural Effusion, LO: Lung Opacity.) Best quality performance in bold.

Method	RadGraph Metrics						NLP Metrics				Classification F1			
	Entity			Relation			B-1	MET.	R_L	SPICE	At	Ed	PE	LO
	P	R	F1	P	R	F1								
RATCHET	–	–	–	–	–	–	**0.18**	**0.09**	0.19	**0.09**	0.38	0.27	0.47	0.17
RATCHET → RadGraph Benchmark	0.45	0.16	0.24	0.11	0.03	0.05	–	–	–	–	–	–	–	–
vanilla RadGraphFormer	0.58	0.47	0.52	0.35	0.19	0.25	0.14	0.08	**0.21**	0.08	**0.40**	0.26	**0.82**	**0.24**
Prior-RadGraphFormer	**0.63**	**0.55**	**0.59**	**0.44**	**0.21**	**0.28**	0.07	0.07	0.19	0.08	0.39	**0.29**	0.81	0.22

to differences in the validation set used in our experiments compared to the original paper. Table 1 shows that our radiology graphs directly generated from CXR images can be properly transformed into free-text reports and pathologies classification results. They have reasonable performance in the corresponding evaluation and demonstrate the utility of a graph-based representation in the context of a clinical study.

From Table 1 it is evident that vanilla RadGraphFormer exhibits better performance with respect to BLEU-1 score and F1 score of several pathologies as compared to Prior-RadGraphFormer. This disparity in performance can be explained by a more varied set of entities in radiology graphs generated by vanilla RadGraphFormer. Also, it is worth noting that the NLP metrics of RATCHET are generally better than our methods. However, higher NLP metrics may not indicate better clinical usefulness [20]. One example is shown in Table 2. Qualitative results can be seen in Fig. 2.

Table 2. Example of evaluating clinical reports with NLP metrics. In this case, our report exhibits superior clinical accuracy compared to RATCHET, but NLP metrics associated with it are significantly lower.

Report section	NLP Metrics				Classification
	B-1	MET	R_L	SPICE	PE
Ground Truth: As compared to the previous radiograph, small left and moderate layering right pleural effusions have increased in size	–	–	–	–	Positive
RATCHET: As compared to the previous radiograph, the patient has been intubated	0.36	0.19	0.47	0.24	Negative
Prior&Vanilla-RadGraphFormer: There is enlarged effusion located at bilateral of pleural	0.07	0.06	0.13	0.12	Positive

Table 3. Ablation studies on the validation set. (AECS: additional entity class supervision)

Method	Entity			Relation		
	Precision	Recall	F1-score	Precision	Recall	F1-score
vanilla RadGraphFormer	0.58	0.47	0.52	0.35	0.19	0.25
Prior-RadGraphFormer w/o PKG	0.61	0.50	0.55	0.39	0.17	0.23
Prior-RadGraphFormer w/o AECS	0.63	0.49	0.55	0.43	0.17	0.24
Prior-RadGraphFormer	**0.63**	**0.55**	**0.59**	**0.44**	**0.21**	**0.28**

4.3 Ablation Studies

In our ablation studies, we aim to investigate two key aspects. Firstly, we aim to determine whether the integration of PKG truly improves performance or whether it is the graph transformers that enhance the model. Secondly, we aim to explore the impact of incorporating additional entity class supervision on entity and relation metrics.

To assess the impact of PKG integration, we simply use nodes features and edges features after graph transformers without assimilation to generate classification outputs. The results presented in Table 3 demonstrate that Prior-RadGraphFormer without PKG is inferior to Prior-RadGraphFormer, thereby highlighting the importance of incorporating PKG as priors in Prior-RadGraph-Former.

As for the influence of incorporating additional entity class supervision in the assimilation step, we conduct a comparative experiment by removing it and analyzing the resulting performance. The experimental results displayed in Table 3 show that the impact is positive.

4.4 Limitations and Outlook

In this work, we evaluated the performance without directly using the graph itself. In the future work we plan to investigate additional graph-specific metrics [21,26] to provide a more comprehensive evaluation. Furthermore, we acknowledge that the number of baselines is limited and aim to explore more comparative models in our ongoing research. Additionally, the inclusion of PKG increases the training time. In future work, we plan to optimize the integration of PKG to accelerate the training without compromising performance.

5 Conclusion

In this paper, we propose Prior-RadGraphFormer, a novel detection-free method that generates radiology graphs directly from CXR images. The model incorporates prior knowledge in the form of probabilistic knowledge graphs that capture the statistical relationship between anatomies and observations. We demonstrate the effectiveness of Prior-RadGraphFormer in the CXR-image-to-radiology-graph task. Moreover, we show that generated radiology graphs are useful for downstream tasks such as free-text reports generation and multi-label classification of pathologies. Our findings offer a promising direction for automatically generating radiology graphs from CXR images. We pave the way to automate the classification of fine-grained clinical findings structured in a radiology graph that can be applied to generate reports while allowing for the assessment of clinical correctness.

Acknowledgements. The authors gratefully acknowledge the financial support by the Federal Ministry of Education and Research of Germany (BMBF) under project DIVA (FKZ 13GW0469C). Kamilia Zaripova was partially supported by the Linde & Munich Data Science Institute, Technical University of Munich Ph.D. Fellowship.

References

1. Cao, Y., Cui, L., Zhang, L., Yu, F., Li, Z., Xu, Y.: MMTN: multi-modal memory transformer network for image-report consistent medical report generation. In: Proceedings of the AAAI Conference on Artificial Intelligence, vol. 37, pp. 277–285 (2023)
2. Carion, N., Massa, F., Synnaeve, G., Usunier, N., Kirillov, A., Zagoruyko, S.: End-to-end object detection with transformers. In: Vedaldi, A., Bischof, H., Brox, T., Frahm, J.-M. (eds.) ECCV 2020. LNCS, vol. 12346, pp. 213–229. Springer, Cham (2020). https://doi.org/10.1007/978-3-030-58452-8_13
3. Cicero, M.D., et al.: Training and validating a deep convolutional neural network for computer-aided detection and classification of abnormalities on frontal chest radiographs. Invest. Radiol. **52**, 281–287 (2017)
4. Goldberger, A.L., et al.: Physiobank, physiotoolkit, and physionet: components of a new research resource for complex physiologic signals. Circulation (2000)
5. Good, I.J.: Rational decisions. J. Roy. Stat. Soc.: Ser. B (Methodol.) **14**(1), 107–114 (1952)
6. Hou, B., Kaissis, G., Summers, R.M., Kainz, B.: RATCHET: medical transformer for chest X-ray diagnosis and reporting. In: de Bruijne, M., et al. (eds.) MICCAI 2021. LNCS, vol. 12907, pp. 293–303. Springer, Cham (2021). https://doi.org/10.1007/978-3-030-87234-2_28
7. Huang, G., Liu, Z., Pleiss, G., Van Der Maaten, L., Weinberger, K.: Convolutional networks with dense connectivity. IEEE Trans. Pattern Anal. Mach. Intell. (2019)
8. Irvin, J., et al.: Chexpert: a large chest radiograph dataset with uncertainty labels and expert comparison. In: Proceedings of the AAAI Conference on Artificial Intelligence, vol. 33, pp. 590–597 (2019)
9. Jain, S., et al.: Radgraph: extracting clinical entities and relations from radiology reports. arXiv preprint arXiv:2106.14463 (2021)
10. Johnson, A., et al.: Mimic-cxr-jpg - chest radiographs with structured labels (version 2.0.0) (2019)
11. Johnson, A.E., et al.: Mimic-cxr-jpg, a large publicly available database of labeled chest radiographs. arXiv preprint arXiv:1901.07042 (2019)
12. Kaur, N., Mittal, A.: RadioBERT: a deep learning-based system for medical report generation from chest x-ray images using contextual embeddings. J. Biomed. Inform. **135**, 104220 (2022)

13. Keicher, M., Mullakaeva, K., Czempiel, T., Mach, K., Khakzar, A., Navab, N.: Few-shot structured radiology report generation using natural language prompts. arXiv preprint arXiv:2203.15723 (2022)
14. Li, R., Zhang, S., He, X.: SGTR: end-to-end scene graph generation with transformer. In: Proceedings of the IEEE/CVF Conference on Computer Vision and Pattern Recognition (CVPR), pp. 19486–19496 (2022)
15. Lin, T.Y., Goyal, P., Girshick, R., He, K., Dollár, P.: Focal loss for dense object detection. In: Proceedings of the IEEE International Conference on Computer Vision, pp. 2980–2988 (2017)
16. Loshchilov, I., Hutter, F.: Decoupled weight decay regularization. In: International Conference on Learning Representations (2017)
17. Lu, C., Krishna, R., Bernstein, M., Fei-Fei, L.: Visual relationship detection with language priors. In: Leibe, B., Matas, J., Sebe, N., Welling, M. (eds.) ECCV 2016. LNCS, vol. 9905, pp. 852–869. Springer, Cham (2016). https://doi.org/10.1007/978-3-319-46448-0_51
18. Lu, Y., et al.: Context-aware scene graph generation with seq2seq transformers. In: Proceedings of the IEEE/CVF International Conference on Computer Vision, pp. 15931–15941 (2021)
19. Ma, Y., Lv, W.: Identification of pneumonia in chest x-ray image based on transformer. J. Antennas Propag. (2022)
20. Pino, P., Parra, D., Besa, C., Lagos, C.: Clinically correct report generation from chest x-rays using templates. In: Lian, C., Cao, X., Rekik, I., Xu, X., Yan, P. (eds.) MLMI 2021. LNCS, vol. 12966, pp. 654–663. Springer, Cham (2021). https://doi.org/10.1007/978-3-030-87589-3_67
21. Rolínek, M., Swoboda, P., Zietlow, D., Paulus, A., Musil, V., Martius, G.: Deep graph matching via blackbox differentiation of combinatorial solvers. In: Vedaldi, A., Bischof, H., Brox, T., Frahm, J.-M. (eds.) ECCV 2020. LNCS, vol. 12373, pp. 407–424. Springer, Cham (2020). https://doi.org/10.1007/978-3-030-58604-1_25
22. Sharifzadeh, S., Baharlou, S.M., Tresp, V.: Classification by attention: scene graph classification with prior knowledge. In: Proceedings of the AAAI Conference on Artificial Intelligence, vol. 35, pp. 5025–5033 (2021)
23. Shin, H.C., Roberts, K., Lu, L., Demner-Fushman, D., Yao, J., Summers, R.M.: Learning to read chest x-rays: recurrent neural cascade model for automated image annotation. In: Proceedings of the IEEE Conference on Computer Vision and Pattern Recognition, pp. 2497–2506 (2016)
24. Shit, S., et al.: Relationformer: a unified framework for image-to-graph generation. In: Avidan, S., Brostow, G., Cissé, M., Farinella, G.M., Hassner, T. (eds.) ECCV 2022, Part XXXVII. LNCS, vol. 13697, pp. 422–439. Springer, Cham (2022). https://doi.org/10.1007/978-3-031-19836-6_24
25. Wang, X., Peng, Y., Lu, L., Lu, Z., Summers, R.M.: Tienet: text-image embedding network for common thorax disease classification and reporting in chest x-rays. In: Proceedings of the IEEE Conference on Computer Vision and Pattern Recognition, pp. 9049–9058 (2018)
26. Wills, P., Meyer, F.G.: Metrics for graph comparison: a practitioner's guide. PLoS ONE 15(2), e0228728 (2020)
27. Wu, J.T., et al.: Chest imagenome dataset for clinical reasoning. arXiv preprint arXiv:2108.00316 (2021)
28. Yu, F., et al.: Evaluating progress in automatic chest x-ray radiology report generation. medRxiv, 2022-08 (2022)
29. Yun, S., Jeong, M., Kim, R., Kang, J., Kim, H.J.: Graph transformer networks. In: Advances in Neural Information Processing Systems, vol. 32 (2019)

A Comparative Study
of Population-Graph Construction
Methods and Graph Neural Networks
for Brain Age Regression

Kyriaki-Margarita Bintsi[1](✉), Tamara T. Mueller[2], Sophie Starck[2],
Vasileios Baltatzis[1,3], Alexander Hammers[3], and Daniel Rueckert[1,2]

[1] BioMedIA, Department of Computing, Imperial College London, London, UK
m.bintsi19@imperial.ac.uk
[2] Lab for AI in Medicine and Healthcare, Faculty of Informatics,
Technical University of Munich, Munich, Germany
[3] Biomedical Engineering and Imaging Sciences, King's College London, London, UK

Abstract. The difference between the chronological and biological brain
age of a subject can be an important biomarker for neurodegenerative
diseases, thus brain age estimation can be crucial in clinical settings.
One way to incorporate multimodal information into this estimation is
through population graphs, which combine various types of imaging data
and capture the associations among individuals within a population. In
medical imaging, population graphs have demonstrated promising results,
mostly for classification tasks. In most cases, the graph structure is pre-
defined and remains static during training. However, extracting population graphs is a non-trivial task and can significantly impact the performance of Graph Neural Networks (GNNs), which are sensitive to the
graph structure. In this work, we highlight the importance of a meaningful graph construction and experiment with different population-graph
construction methods and their effect on GNN performance on brain age
estimation. We use the homophily metric and graph visualizations to
gain valuable quantitative and qualitative insights on the extracted graph
structures. For the experimental evaluation, we leverage the UK Biobank
dataset, which offers many imaging and non-imaging phenotypes. Our
results indicate that architectures highly sensitive to the graph structure,
such as Graph Convolutional Network (GCN) and Graph Attention Network (GAT), struggle with low homophily graphs, while other architectures, such as GraphSage and Chebyshev, are more robust across different
homophily ratios. We conclude that static graph construction approaches
are potentially insufficient for the task of brain age estimation and make
recommendations for alternative research directions.

Keywords: Brain age regression · Population graphs · Graph Neural
Networks

1   Introduction

Alzheimer's disease [9], Parkinson's disease [24], and schizophrenia [18], among other neurodegenerative diseases, cause an atypically accelerated aging process in the brain. Consequently, the difference between an individual's biological brain age and their chronological age can act as an indicator of deviation from the normal aging trajectory [2], and potentially serve as a significant biomarker for neurodegenerative diseases [7,13].

Graph Neural Networks (GNNs) have been recently explored for medical tasks, since graphs can provide an inherent way of combining multi-modal information, and have demonstrated improved performance in comparison to graph-agnostic deep learning models [1,23]. In most cases, the whole population is represented in a graph, namely population-graph, and the structure of the graph is pre-decided and remains static throughout the training. However, since the structure is not given but has to be inferred from the data, there are multiple ways that a graph can be constructed, which will possibly lead to different levels of performance. Every subject of the cohort is a node of the graph, with the imaging information usually allocated as node features. The interactions and relationships among the subjects of the cohort are represented by the edges. Two nodes are chosen to be connected based on a distance measure, or similarity score, usually based on the non-imaging phenotypes. The population graph is used as input to a GNN for the task of node prediction, most commonly node classification.

Brain MRI data have been widely used in population graphs in combination with GNNs. Parisot et al. were the first to propose a way to construct a static population graph using a similarity score that takes into account both imaging and non-imaging data [23]. Many works were based on this work and extended the method to other applications [16,28]. When it comes to brain age regression, which is an inherently more complicated task than classification, there is little ongoing work using population graphs. To our knowledge, only Stankevičiūtė et al. [25] worked on the construction of the static population graph using the non-imaging information, but the predictive performance was relatively low.

In all of the cases mentioned above, the graph is chosen based on some criterion before training and remains static, thus it cannot be changed. However, since there are multiple ways that it can be constructed, there is no way to ensure that the final structure will be optimized for the task. This problem has been identified in the literature and multiple metrics have been described in order to evaluate the final graph structure [21,29], with homophily being the one most commonly used. A graph is considered homophilic, when nodes are connected to nodes with the same class label, and hence similar node features. Else, it is characterized as heterophilic. It has been found that some GNNs, such as Graph Convolutional Network (GCN) [17], and Graph Attention Network (GAT) [26], are very sensitive to the graph structure, and a meaningless graph along with these models can perform worse than a graph-agnostic model. On the other hand, other models, such as GraphSage [14] and Chebyshev [10], are more resilient to the graph structure, and their performance is not affected as much by a heterophilic graph [30].

In this work, we implement and evaluate the performance of different static graph construction methods for the task of brain age regression. We test their performance on the UK Biobank (UKBB), which offers a variety of both imaging and non-imaging phenotypes. The extracted population graphs are used along with the most popular GNNs, and more specifically GCN [17] and GAT [26], which are architectures highly sensitive to the graph structure, as well as GraphSage [14] and Chebyshev [10], which have been found to be resilient to the population-graph structure. The quantitative results and the visualization of the graphs allow us to draw conclusions and make suggestions regarding the use of static graphs for brain age regression. The code is available on GitHub at: https://github.com/bintsi/brain-age-population-graphs.

2   Methods

We have a dataset consisting of a set of N subjects, each described by M features. This dataset can be represented as $\mathbf{X} = [\mathbf{x}_1, \ldots, \mathbf{x}_N] \in \mathbb{R}^{N \times M}$, where $\mathbf{x}_i$ represents the feature vector for the i-th subject. Additionally, each subject has a label denoted by $\mathbf{y} \in \mathbb{R}^N$. Every subject i is also characterized by a set of K non-imaging phenotypes $\mathbf{q}_i \in \mathbb{R}^K$.

To establish the relationship between subjects, we introduce a population graph, denoted as $\mathcal{G} = \{\mathcal{V}, \mathcal{E}\}$. The graph consists of two components: $\mathcal{V}$ represents the set of nodes, where each subject corresponds to a unique node, and $\mathcal{E}$ represents the set of edges that define the connectivity between nodes.

In this paper, we explore four different graph construction approaches, which are described in detail below. In all of the graph construction methods, the node features consist of the imaging features described in the Dataset Sect. 3.1.

No Edges. As the baseline, we consider a graph with no edges among the nodes. This is the equivalent of traditional machine learning, where the node features act as the features used as input in the model. The machine learning model in this case is a Multi-Layer Perceptron (MLP). We refer to this approach as *"No edges"*.

Random Graph. With the next approach, we want to explore whether the way of constructing the graph plays an important role on the performance of the GNN. Thus, we build a random Erdos-Renyi graph [11], where a random number of nodes is chosen as neighbors for every node.

Clinical Similarity Score (Stankevičiūtė et al.). An approach of creating a population graph specifically for the task of brain age regression was proposed by [25], and it does not include the imaging features at all in the extraction of the edges. Instead, the edges are decided only from the non-imaging information. More specifically, the similarity function for two subjects i and j is given by:

$$sim(i, j) = \frac{1}{K} \sum_{k=1}^{K} \mathbf{1}[q_{ik} = q_{jk}], \tag{1}$$

where q_{ik} is the value of the k-th non-imaging phenotype for the k-th subject, and $\mathbf{1}$ is the Kronecker delta function. Intuitively, the Kronecker delta function will only return 1 if the values of a particular non-imaging phenotype of two subjects match. Two nodes i and j are connected if $sim(i,j) \geq \mu$, with μ being a similarity threshold decided empirically.

Similarity Score (Parisot et al.). In most of the related works in the literature, the way of creating the adjacency matrix W was originally suggested by [23] and it is calculated by:

$$W(i,j) = Sim(\mathbf{x}_i, \mathbf{x}_j) \sum_{k=1}^{K} \gamma(q_{ik}, q_{jk}), \qquad (2)$$

where $Sim(\mathbf{x}_i, \mathbf{x}_j)$ is a similarity measure between the node features $(\mathbf{x}_i, \mathbf{x}_j)$ of the subjects i and j, in our case cosine similarity, $\gamma(\cdot, \cdot)$ is the distance of the non-imaging phenotypes between the nodes, and q_{ik} is the value of the k-th non-imaging phenotype for the i-th subject.

The two terms in Eq. 2 indicate that both imaging, and non-imaging information are taken into account for the extraction of the edges. The second term is similar to the term $sim(i,j)$ of Eq. 1.

The computation of $\gamma(\cdot, \cdot)$ is different for continuous and categorical features. For categorical data, $\gamma(\cdot, \cdot)$ is defined as the Kronecker delta function $\mathbf{1}$, as before. For continuous data, $\gamma(\cdot, \cdot)$ is defined as a unit-step function with respect to a specific threshold θ. Intuitively, this means that the output of the γ function will be 1 in case the continuous phenotypes of the two nodes are similar enough.

kNN Graph. The last two graph construction approaches are based on the Nearest Neighbors (NN) algorithm. We connect the edges based on a distance function, in this case, cosine similarity. Each node is connected to its 5 closest neighbors.

In the first approach, we use the neuroimaging information for the node features, and we also use the node features in order to estimate the distances of the nodes. We refer to this approach as *"kNN (imaging)"*.

Similarly to before, in the second approach, we use the cosine similarity of a set of features in order to extract the graph structure. The node features incorporate the imaging information as before. The difference here is that we estimate the cosine similarity of the non-imaging phenotypes of the subjects with the purpose of finding the 5 closest neighbors. We refer to this approach as *"kNN (non-imaging)"*.

Finally, we use all the available phenotypes, i.e. both the imaging, and non-imaging information, to connect the nodes, again using cosine similarity. We refer to this approach as *"kNN (all phenotypes)"*.

Table 1. Performance of the different graph construction methods along with various GNNs on the test set. For every population graph, the homophily ratio is estimated. The baseline MLP (*No edges*) has a performance of MAE = 3.73 years and R^2 score of 0.56. Best performance for each GNN model is highlighted in bold.

Graph Construction	GCN	GraphSAGE	GAT	Chebyshev	Homophily
	MAE (years)				
Random graph	5.19	3.72	5.38	3.83	0.7495
Parisot [23]	4.21	3.74	4.35	3.77	0.7899
Stankevičiūtė [25]	4.61	3.73	4.90	3.77	0.7743
kNN (imaging)	**3.89**	3.77	4.09	3.75	0.8259
kNN (non-imaging)	4.76	**3.68**	4.98	**3.72**	0.7796
kNN (all phenotypes)	3.93	3.76	**4.07**	3.73	0.8191
	R^2 score				
Random graph	0.26	0.59	0.2	0.54	
Parisot [23]	0.49	0.59	0.47	0.55	
Stankevičiūtė [25]	0.4	0.59	0.3	0.58	
kNN (imaging)	**0.56**	0.58	0.51	0.59	
kNN (non-imaging)	0.38	**0.59**	0.3	0.55	
kNN (all phenotypes)	0.56	0.58	**0.53**	**0.6**	

3   Experiments

3.1   Dataset

For the experiments of the comparative study we use the UK Biobank (UKBB) [3], which not only offers an extensive range of vital organ images, including brain scans, but also contains a diverse collection of non-imaging information such as demographics, biomedical data, lifestyle factors, and cognitive performance measurements. Consequently, it is exceptionally well-suited for brain age estimation tasks that require integrating both imaging and non-imaging information.

To identify the most important factors influencing brain age in the UKBB, we leverage the work of [6] and select 68 neuroimaging phenotypes and 20 non-imaging phenotypes that were found to be the most relevant to brain age in UKBB. The neuroimaging features are obtained directly from the UKBB and include measurements derived from both structural MRI and diffusion-weighted MRI. All phenotypes are standardized to a normalized range between 0 and 1.

The study focuses on individuals aged 47 to 81 years. We include only those subjects who have the necessary phenotypes available, resulting in a group of approximately 6500 subjects. The dataset is split into three parts: 75% for training, 5% for validation, and 20% for testing.

3.2 Results

The graph construction methods described in the previous section are leveraged by four GNN architectures and evaluated on the task of brain age regression. For every experiment, both spectral and spatial methods, and more specifically, GCN [17], GAT [26], GraphSage [14], and Chebyshev [10], are used. The results for all combinations of graph construction methods and GNN architectures are shown in Table 1.

As a baseline, we use an MLP (*No edges*) that achieves a MAE of 3.73 years and a R^2 score of 0.56. For the case of GCN and GAT, we notice that even though all of the graph construction methods manage to extract a graph that outperforms the random graph, none of them manages to outperform the simple MLP. The methods that used only the imaging information, i.e. the *kNN graph (imaging)*, or all of the phenotypes, such as *Parisot et al.* [23], extract graphs that when used as input to the GCN or the GAT models, they perform better (MAE of around 4 years) compared to the ones that only use the non-imaging features for the extraction of the graph, such as *kNN graph (non-imaging)* (MAE of 4.5–5 years). The GraphSAGE and Chebyshev models perform on similar levels regardless of the graph construction methods, with a MAE similar to the MAE given by the MLP. The graph construction method that performs the best though, is the *kNN graph (non-imaging)* along with the GraphSAGE, with a MAE of 3.68 years and a R^2 score of 0.59.

Finally, we estimate the homophily ratio of the population graphs extracted from the different construction methods based on [22] (Table 1). The most homophilic graphs tend to be the ones that use only imaging or a combination of imaging and non-imaging information for the extraction of the edges.

3.3 Visualizations

Visualizing the extracted static graphs, can provide more insights about the performance of the different GNN architectures. Therefore, we visualize the different population graphs colored based on the age of the subjects in Fig. 1. The various graph construction methods result in very different graph structures. More specifically, the graph construction methods that take into account the imaging phenotypes, either alone or in a combination with the non-imaging ones, provide graphs that are more meaningful, as subjects of similar ages are closer to each other. On the contrary, methods such as Stankevičiūtė et al. [25], create graphs that are similar to the random graph, with the neighborhoods not being very informative regarding the age of the subjects.

3.4 Implementation Details

The static graphs extracted are chosen to be sparse for complexity reasons. The hyperparameters of the different construction methods were selected in such a way that all the extracted graphs have about 40000 to 50000 edges. Regarding the model hyperparameters, every GNN contains a graph convolutional layer

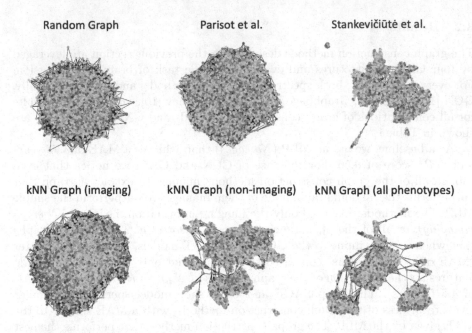

Fig. 1. Visualizations of the population graphs extracted from the different graph construction methods. The nodes are colored based on the age of the subject. Colder colors (i.e. blue) indicate subjects with an older age, while warmer colors (i.e. red) correspond to younger subjects. (Color figure online)

consisting of 512 units, followed by a fully connected layer with 128 units, prior to the regression layer. ReLU activation is chosen. The number of layers and their dimensions are determined by conducting a hyperparameter search based on validation performance. During training, the networks are optimized using the AdamW optimizer [19] with a learning rate of 0.001 for 150 epochs, with the best model being saved. For the similarity threshold we use $\mu = 18$ and for the unit-step function threshold, we use $\theta = 0.1$. Both hyperparameters are selected based on validation performance and sparsity requirements. The implementation utilizes PyTorch Geometric [12] and a Titan RTX GPU.

4   Discussion and Conclusions

In this work, we implement and evaluate static population graphs that are commonly used in the literature for other medical tasks, for brain age regression. We use the extracted graphs along with a number of popular GNN models, namely GCN, GAT, GraphSAGE, Chebyshev in order to get insights about the behavior of both the extraction methods, as well as the performance of the different models. By visualizing the graphs and estimating their homophily, we can provide further intuition in why the different static graphs do not work as expected and we highlight the problem that extracting static graphs from the data is not straightforward and possibly not suitable for brain age regression.

The reported results in Table 1 indicate that the GCN and GAT are highly sensitive to the graph structure, which is in agreement with the relevant literature [20,30]. It is clear that the graph construction methods that provide graphs of higher homophily, lead to better performance for GCN and GAT compared to the ones that provide a random-like population graph.

On the contrary, GraphSAGE and Chebyshev are not negatively affected by a graph structure with low homophily. This is because GraphSAGE encodes separately the node's encodings and the neighbors' encodings, which in our case are dissimilar, and hence it is affected less by the graph structure. When it comes to Chebyshev, the model is able to aggregate information from k hops in one layer, while the other GNNs achieve this through multiple layers. Being able to get information from higher order neighborhoods allows the model to find more relevant features, which would not be possible in the 1-hop neighborhood as this is highly heterophilic. But even these GNNs, that are more resilient to the graph structure, cannot distinctly outperform the MLP. This behavior is expected as, according to the relevant literature, these models perform similarly to a MLP, and they can outperform it only after a specific homophily threshold [30]. What is also observed is that the *kNN (non-imaging)* works slightly better than the MLP both for the GraphSAGE and the Chebyshev, probably because the models were able to capture the information added in the edges.

To further explore the behavior of the graph construction methods and the population-graph they produce, we calculate the homophily ratio and we visualize the graphs. The graphs that are more homophilic perform better along with the models that are more sensitive to the graph structure, such as the GCN and the GAT. We note here that all of the homophily ratios are higher than expected compared to the homophily reported in classification tasks, since we would expect that the random graph would have a homophily of 0.5. This is possibly because of the implementation of homophily for regression, as well as due to the imbalanced nature of the dataset. However, the trend is very clear and in agreement with the graph visualizations and the performance of the models.

All in all, the extraction of static population graphs for brain age regression, and in general for medical tasks for which the graph is not given, does not look very promising. In our opinion, there are multiple directions that should be explored. Firstly, one approach could be to learn the edges of the graph along with the training of the GNN, which allows the extraction of an optimized graph for the specific task at-hand. There is some ongoing work on this adaptive graph learning [5,8,15,27], but more focus should be given. In addition, the creation of GNN models for graphs with high heterophily [29,30], or the exploitation of graph rewiring techniques that could make the existing GNN work better on heterophilic graphs [4], have proved to be useful for classification tasks. In the case of medical tasks, it might also be beneficial to incorporate the existing medical insights along with the above. It is also important to make the models and the graphs interpretable, as interpretability can be vital in healthcare, and it is something that is not currently widely explored in heterophilic graphs [29]. Last but not least, introducing metrics that evaluate population graphs is of high importance, since this can help us understand better the structure of the graph. Even though there is some ongoing work when it comes to node classification [20,29], metrics regarding node regression have only recently started to being explored [22].

Acknowledgements. KMB would like to acknowledge funding from the EPSRC Centre for Doctoral Training in Medical Imaging (EP/L015226/1).

References

1. Ahmedt-Aristizabal, D., Armin, M.A., Denman, S., Fookes, C., Petersson, L.: Graph-based deep learning for medical diagnosis and analysis: past, present and future. Sensors **21**(14), 4758 (2021)
2. Alam, S.B., Nakano, R., Kamiura, N., Kobashi, S.: Morphological changes of aging brain structure in MRI analysis. In: 2014 Joint 7th International Conference on Soft Computing and Intelligent Systems (SCIS) and 15th International Symposium on Advanced Intelligent Systems (ISIS), pp. 683–687. IEEE (2014)
3. Alfaro-Almagro, F., et al.: Image processing and quality control for the first 10,000 brain imaging datasets from UK Biobank. NeuroImage **166**, 400–424 (2018)
4. Bi, W., Du, L., Fu, Q., Wang, Y., Han, S., Zhang, D.: Make heterophily graphs better fit GNN: a graph rewiring approach. arXiv preprint arXiv:2209.08264 (2022)
5. Bintsi, K.M., Baltatzis, V., Potamias, R.A., Hammers, A., Rueckert, D.: Multimodal brain age estimation using interpretable adaptive population-graph learning. In: Greenspan, H., et al. (eds.) MICCAI 2023. LNCS, vol. 14227, pp. 195–204. Springer, Cham (2023). https://doi.org/10.1007/978-3-031-43993-3_19
6. Cole, J.H.: Multimodality neuroimaging brain-age in UK biobank: relationship to biomedical, lifestyle, and cognitive factors. Neurobiol. Aging **92**, 34–42 (2020)
7. Cole, J.H., et al.: Predicting brain age with deep learning from raw imaging data results in a reliable and heritable biomarker. Neuroimage **163**, 115–124 (2017)
8. Cosmo, L., Kazi, A., Ahmadi, S.-A., Navab, N., Bronstein, M.: Latent-graph learning for disease prediction. In: Martel, A.L., et al. (eds.) MICCAI 2020. LNCS, vol. 12262, pp. 643–653. Springer, Cham (2020). https://doi.org/10.1007/978-3-030-59713-9_62
9. Davatzikos, C., Bhatt, P., Shaw, L.M., Batmanghelich, K.N., Trojanowski, J.Q.: Neurobiol. AgingPrediction of mci to ad conversion, via MRI, CSF biomarkers, and pattern classification **32**(12), 2322-e19 (2011)
10. Defferrard, M., Bresson, X., Vandergheynst, P.: Convolutional neural networks on graphs with fast localized spectral filtering. In: Advances in Neural Information Processing Systems, vol. 29 (2016)
11. Erdős, P., Rényi, A., et al.: On the evolution of random graphs. Publ. Math. Inst. Hung. Acad. Sci. **5**(1), 17–60 (1960)
12. Fey, M., Lenssen, J.E.: Fast graph representation learning with PyTorch geometric. In: ICLR Workshop on Representation Learning on Graphs and Manifolds (2019)
13. Franke, K., Gaser, C.: Ten years of brainage as a neuroimaging biomarker of brain aging: what insights have we gained? Front. Neurol. 789 (2019)
14. Hamilton, W., Ying, Z., Leskovec, J.: Inductive representation learning on large graphs. In: Advances in Neural Information Processing Systems, vol. 30 (2017)

15. Kazi, A., Cosmo, L., Ahmadi, S.A., Navab, N., Bronstein, M.: Differentiable graph module (DGM) for graph convolutional networks. IEEE Trans. Pattern Anal. Mach. Intell. (2022)

16. Kazi, A., et al.: InceptionGCN: receptive field aware graph convolutional network for disease prediction. In: Chung, A.C.S., Gee, J.C., Yushkevich, P.A., Bao, S. (eds.) IPMI 2019. LNCS, vol. 11492, pp. 73–85. Springer, Cham (2019). https://doi.org/10.1007/978-3-030-20351-1_6

17. Kipf, T.N., Welling, M.: Semi-supervised classification with graph convolutional networks. arXiv preprint arXiv:1609.02907 (2016)

18. Koutsouleris, N., et al.: Accelerated brain aging in schizophrenia and beyond: a neuroanatomical marker of psychiatric disorders. Schizophr. Bull. 40(5), 1140–1153 (2014)

19. Loshchilov, I., Hutter, F.: Decoupled weight decay regularization. arXiv preprint arXiv:1711.05101 (2017)

20. Luan, S., Hua, C., Lu, Q., Zhu, J., Chang, X.W., Precup, D.: When do we need GNN for node classification? arXiv preprint arXiv:2210.16979 (2022)

21. Ma, Y., Liu, X., Shah, N., Tang, J.: Is homophily a necessity for graph neural networks? arXiv preprint arXiv:2106.06134 (2021)

22. Mueller, T., Starck, S., Feiner, L.F., Bintsi, K.M., Rueckert, D., Kaissis, G.: Extended graph assessment metrics for regression and weighted graphs. arXiv preprint (2023)

23. Parisot, S., et al.: Disease prediction using graph convolutional networks: application to autism spectrum disorder and Alzheimer's disease. Med. Image Anal. 48, 117–130 (2018)

24. Reeve, A., Simcox, E., Turnbull, D.: Ageing and Parkinson's disease: why is advancing age the biggest risk factor? Ageing Res. Rev. 14, 19–30 (2014)

25. Stankeviciute, K., Azevedo, T., Campbell, A., Bethlehem, R., Lio, P.: Population graph GNNs for brain age prediction. In: ICML Workshop on Graph Representation Learning and Beyond (GRL+), pp. 17–83 (2020)

26. Veličković, P., Cucurull, G., Casanova, A., Romero, A., Lio, P., Bengio, Y.: Graph attention networks. arXiv preprint arXiv:1710.10903 (2017)

27. Wei, S., Zhao, Y.: Graph learning: a comprehensive survey and future directions. arXiv preprint arXiv:2212.08966 (2022)

28. Zhao, X., Zhou, F., Ou-Yang, L., Wang, T., Lei, B.: Graph convolutional network analysis for mild cognitive impairment prediction. In: 2019 IEEE 16th International Symposium on Biomedical Imaging (ISBI 2019), pp. 1598–1601. IEEE (2019)

29. Zheng, X., Liu, Y., Pan, S., Zhang, M., Jin, D., Yu, P.S.: Graph neural networks for graphs with heterophily: a survey. arXiv preprint arXiv:2202.07082 (2022)

30. Zhu, J., Yan, Y., Zhao, L., Heimann, M., Akoglu, L., Koutra, D.: Beyond homophily in graph neural networks: current limitations and effective designs. In: Advances in Neural Information Processing Systems, vol. 33, pp. 7793–7804 (2020)

Self Supervised Multi-view Graph Representation Learning in Digital Pathology

Vishwesh Ramanathan[1,2]($\boxtimes$) and Anne L. Martel[1,2]

[1] Physical Sciences, Sunnybrook Research Institute, Toronto, Canada
[2] Department of Medical Biophysics, University of Toronto, Toronto, Canada
`vishwesh.ramanathan@mail.utoronto.ca`

Abstract. Graph Neural Networks (GNNs) hold great promise for solving many challenges in digital pathology by leveraging the rich relationships between cells and tissues in histology images. However, the shortage of annotated data in digital pathology presents a significant challenge for training GNNs. To address this, self-supervision can be used to enable models to learn from data by capturing rich structures and relationships without requiring annotations. Inspired by pathologists who take multiple views of a histology slide under a microscope for exhaustive analysis, we propose a novel methodology for graph representation learning using self-supervision. Our methodology leverages multiple graph views constructed from a given histology image to capture diverse information. We maximize mutual information across nodes and graph representations of different graph views, resulting in a comprehensive graph representation. We showcase the efficacy of our methodology on the BRACS dataset where our algorithm generates superior representations compared to other self-supervised graph representation learning algorithms and comes close to pathologists and supervised learning algorithms. The code and pre-trained weights are shared on github at https://github.com/Vishwesh4/Multiview-GRL

Keywords: Self Supervised Learning · Graph Representation Learning · Digital Pathology · Multiple Views

1 Introduction

Digital pathology is a rapidly growing field that employs computerized methods to analyze tissue samples. Graph Neural Networks (GNNs) have demonstrated impressive performances in various problem spaces of digital pathology, such as survival prediction [16], tumor classification [20], grading [2], and more [1]. This is due to their ability to extract information from complex interactions between entities like cells and tissues. Their flexibility in dealing with variable sizes and shapes of samples makes them indispensable tools. However, obtaining a large set of well-annotated data for training these models is very tedious, costly, and time-consuming.

© The Author(s), under exclusive license to Springer Nature Switzerland AG 2024
S.-A. Ahmadi and S. Pereira (Eds.): MICCAI 2023, LNCS 14373, pp. 74–84, 2024.
https://doi.org/10.1007/978-3-031-55088-1_7

Graph representation learning (GRL) using self-supervision aims to capture the underlying structure and relationships in graphs to learn a compact low-dimensional representation without manual annotation. These learned representations can be useful for various downstream tasks such as classification, regression, etc. Self-supervision has already proven effective in Convolutional Neural Networks (CNNs), as demonstrated in [8,22], but there is a need for the same in GNNs. Although self-supervised learning (SSL) has been used for learning node embeddings for entities in digital pathology [6,26] before using them for supervised learning, little has been done on using self-supervised for GRL in this field. Ozen et al. [18] used Graph Contrastive Learning (GCL) [28] to contrast two different graph views of a given cell graph generated by vertex dropping to generate an ROI representation, which was later used for an ROI retrieval downstream task. However, this method uses augmentations that may hamper the topological distribution of biological entities, potentially affecting some downstream tasks. Oscar et al. [21] performed unsupervised analysis on the graph of a whole slide image (WSI) constructed using clusters of cells as nodes spread across the graph. The representation of this graph was learned using InfoGraph [23]. However, the methods mentioned above are restricted to only a single view provided by the cell graph and do not fully utilize the variety of information that a histology image offers. Histology images contain vast information that can be used to create graphs based on biologically relevant concepts.

A single image can have multiple views of graphs, each containing different information, for instance, graph nodes with entities like cells [29], clusters of cells [21], tissues [20], patches [3], superpixels [2], etc. and edges capturing the interactions between them. Pathologists often rely on multiple views to provide a final analysis, such as scanning at higher resolutions for local cellular information and lower resolutions for global tissue microenvironments. We propose a novel methodology that leverages multiple graph views of histology images to learn comprehensive graph representations. Our approach involves maximizing mutual information (MI) between graph representations of one view and node representations of the same and different views. Specifically, we utilize the cell and patch graphs to capture fine-level local cellular information and coarse-level global tissue microenvironments information. This highly adaptable framework can pre-train any GNN encoders with additional graph views without modifying existing user-defined input-output pipelines. Our approach draws inspiration from recent work on Multi-view Graph Representation Learning (MVGRL) [14], but differs in that, MVGRL uses graph augmentation techniques like diffusion [12] to create additional graph views, but these alter the topological distribution of entities, potentially affecting downstream tasks. In contrast, our multiple graph views are biologically inspired, utilizing the information provided by different graph views constructed from histology images. We evaluate our methodology on BRACS [5], a publicly available dataset of tumor regions of interest (TRoIs) for classification downstream task. Our results demonstrate superior performance compared to other self-supervised GRL algorithms and approach the performance of pathologists and supervised learning algorithms.

2 Methods

In our methodology, we perform self-supervised representation learning on graphs constructed from hematoxylin and eosin-stained (H&E) histology images. To account for color variability in the images due to staining, type of scanners used, and slide preparation protocols, we apply stain normalization using the methodology proposed by Vahadane et al. [25]. Subsequently, we construct cell and patch graphs using the stain-normalized images. Finally, we combine information from both graphs by MI maximization between the different graph views.

2.1 Graph Construction

Cell Graph. To construct the nodes for our cell graph, we use Hovernet [13], a nuclear segmentation model pretrained on the Panuke dataset [11] to detect all the nuclei present in the stain-normalized histology image. Graph-level representations are typically obtained by pooling all node representations in a given graph. To prevent oversaturating the graph-level representations, we sample a maximum of 3500 nuclei using farthest sampling [9]. To construct edges, we use K-Nearest Neighbors with 5 neighbors and a threshold based on the Euclidean distance of the nuclei to prevent distant connections. The node feature vectors are then formed by extracting a 72×72 patch around the centroid of each nucleus, resizing it, and passing it through a Resnet34 [15] pre-trained using SimCLR [7] on the BRACS [5] training set. We also consider the class of each nucleus identified by HoverNet and concatenate the one-hot encoding of the class label with the Resnet34-extracted feature vector to form a final node feature vector.

Patch Graph. A patch graph consists of a graph with nodes as patches sampled across the input image. There can be many ways of sampling; for instance, sampling randomly, uniformly, based on texture, using attention networks, etc. For our case, we simply divide the input image into a regular 128×128 grid and extract patches uniformly. These patches form nodes for the patch graph. The edges of the graph are formed based on the adjacency of the surrounding patches, hence forming a lattice graph, fully covering the input image. Finally, we define the node feature vector for each node by passing it through the same SimCLR pre-trained Resnet34 network.

2.2 Graph Representation Learning

Single View (SV). We employ InfoGraph [23], a self-supervised GRL algorithm, to generate graph-level representations for our single graph views i.e. cell graphs and patch graphs. InfoGraph maximizes MI between node-level representations and the final graph-level representation through a contrastive approach. Specifically, MI is maximized for representations coming from the same graph and minimized for representations coming from different graphs. To generate negative samples, the methodology employs a batch-wise approach where one sample is contrasted with the rest of the samples in the batch. This batch-wise generation of negative samples enhances the effectiveness of the algorithm. Info-Graph uses Jensen-Shannon MI estimator [17]. For a given graph from a graph dataset $G \in \mathbb{G}$, let $z_i(G)$ be the ith node embedding, $Z(G)$ be the matrix consisting of embeddings of all nodes in the graph, and $h(G)$ be the graph level embeddings generated by GNN encoder with parameters ϕ, we have

$$I(Z(G), h(G)) = \frac{1}{|G|} \sum_{i \in G} \mathbb{E}_{\mathbb{P}}[-sp(-\mathcal{D}(z_i(x), h(x)))] - \mathbb{E}_{\mathbb{P} \times \tilde{\mathbb{P}}}[sp(\mathcal{D}(z_j(x'), h(x)))]$$

(1)

, where $\mathcal{D}$ is a discriminator parameterized by a neural network with parameters ψ, which outputs a scalar indicating the similarity between two vectors. x is an input graph sample, x' and j is a different graph sample and node sampled from distribution $\tilde{\mathbb{P}} = \mathbb{P}$, which is the empirical probability distribution of the input space. sp denotes softplus function $sp(x) = log(1 + e^x)$. InfoGraph finally maximizes the sum of $I(Z(G), h(G))$ across all the graphs G in the dataset $\mathbb{G}$.

Multi View (MV). We propose a novel methodology that combines information from multiple graph views, specifically cell graph and patch graph, using MI maximization to encourage the generation of a comprehensive representation. Our approach involves enriching the representation generated by the encoder for the patch graph by utilizing information from the cell graph, and vice versa. This results in a multiview representation that captures complementary information from both graph views. Our methodology is illustrated in Fig. 1.

In our approach, following the same procedure as in the previous section, we maximize the MI between the patch node level representation with graph level representation and the same with cell node level representation with graph level representation. For inculcating multiple graph views, we also maximize the MI between cell graph nodes and patch graph level representation and vice versa. We hypothesize that this encourages the encoder to learn representations that have both fine-level topology information such as that obtained from cell graphs, and also coarser global view information obtained from patch graphs. Maximizing the MI between cell graph and patch graph representations forces this effect. Let the two graph views, cell graph, and patch graph for the same image be denoted as $G^c, G^p \in \mathbb{G}$. Let ϕ^p, ϕ^c, be the parameters of encoders for patch graph and cell graph, and ψ^p, ψ^c, be the parameters of the discriminator used in patch graph and cell graph. Using the expression from Eq. 1, we optimize for the following objective function

$$\max_{\psi^p, \psi^c, \phi^p, \phi^c} \sum_{G^c, G^p \in \mathbb{G}} I_{\text{Same Views}} + I_{\text{Different Views}}$$

(2)

where,

$$I_{\text{Same Views}} = I(Z(G^c), h(G^c)) + I(Z(G^p), h(G^p))$$
$$I_{\text{Different Views}} = I(Z(G^p), h(G^c)) + I(Z(G^c), h(G^p))$$

However, the disparity in node numbers between different graph views poses a challenge. For example, the cell graph has an average of 1293 ± 1091 nodes compared to 220 ± 253 nodes for the patch graph in the BRACS training set. This discrepancy can hinder the maximization of mutual information between the graph embedding of the smaller graph and the numerous individual nodes of the larger graph, resulting in suboptimal graph representations. To address this, we adopt a sampling approach to select a subset of nodes from the larger graph (cell graph) to maximize against the graph embedding of the smaller graph (patch graph). Sampling the other way around proved less beneficial due to the already low number of patch graph nodes. Our sampling strategy combines random selection and nodes with the most dissimilar node representation compared to the patch graph representation, determined using a dot product between the representations. In our training process, we employ batch-wise generation of negative samples, contrasting one sample with the rest of the batch using discriminators to calculate individual terms in Eq. 2. For downstream tasks, any of the encoders can be utilized. While the representations can be merged, as shown by Zhou et al. [30], keeping them separate enables users to pretrain encoders using multiple graph views while maintaining their input-output pipeline. Furthermore, certain graph views are time-consuming to create during the inference step, hence our framework offers flexibility and adaptability.

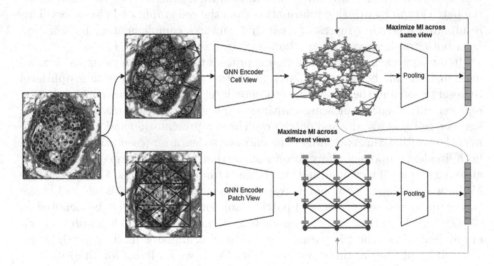

Fig. 1. Proposed methodology for multiview representation learning. The colored arrows indicate the maximization of MI across various cell and patch graph entities.

3 Experiments

Datasets. We evaluated the representations generated by the algorithm on publically available dataset namely Breast Carcinoma Sub-typing [5] (BRACS) for the classification downstream task. BRACS consists of 4391 Tumor regions of interest (TRoIs) from 325 H&E breast carcinoma WSIs, scanned at a resolution of $0.25\,\mu m$/pixel. Each histology image varies highly in its dimensions and is annotated with one of the seven classes: Normal, Benign, Usual Ductal Hyperplasia (UDH), Atypical Ductal Hyperplasia (ADH), Flat Epithelial Atypia (FEA), Ductal Carcinal In Situ (DCIS), and Invasive. To facilitate evaluation and comparison, we use the same dataset (initial version) and WSI splitting scheme (training (3163)/validation (602)/test (626)) as in the paper [20].

Setup. We used Pytorch(v1.12.1) [19] and Pytorch Geometric(v2.1.0) [10] library, and Weights and Biases [4] for logging purposes. We trained a GNN with 4 Graph Isomorphism Network (GIN) [27] layers. We chose GIN encoders due to their high expressive powers for building better graph-level representations. Each GIN layer contains a 2-layer MLP with ReLU activation followed by a batch norm layer. We concatenate the node-level features across different hidden layers for each node and pass through an MLP to obtain the final node features. For graph-level representation, we perform mean pooling on the nodes present at each hidden layer, concatenate across all the layers, and pass through the same MLP to obtain the final feature vector. We did not use sum pooling to prevent learning bias related to the size of the graph. We opted to freeze the encoders and train a linear model on top of the generated feature vectors to evaluate the strength of the representations. To facilitate further experiments, we will make the model weights publicly available. For classification on the BRACS dataset, we utilized logistic regression on the feature vectors to perform 7-way classification and evaluated the model's performance on the test set using weighted F1 score metrics. The encoders were trained for 150 epochs with a batch size 32 using an Adam optimizer and a cyclic learning rate with a triangular scheduler.

4 Results

For our experiments, we evaluated the performances of representations generated by both the cell graph (CG) and patch graph (PG) encoders. We compare our methodology denoted by MV (Multiview) with InfoGraph and Adversarial Graph Contrastive Learning (ADGCL) [24]. ADGCL is a self-supervised GRL algorithm that aims to maximize MI between the encoder-generated representations of the original graph and its augmented version. ADGCL uses an adaptive edge-dropping strategy to generate an augmented graph, removing redundant information compared to its original counterpart. It outperforms InfoGraph on most of the graph datasets and achieves state-of-the-art performances. We also include results from a baseline supervised learning approach for both CG and PG encoders, using the same architecture for comparison, as shown in the row

Table 1. BRACS results on 7-class classification downstream tasks. Values of per-class F1 scores and weighted F1 scores are shown in the form *mean ± std*, expressed in %. The best-performing self-supervised algorithm across both graphs is in bold, and the best-performing supervised algorithm is underlined. Figures for all algorithms in the row *Supervised Algorithms* and *Pathologist Scores* is taken from the paper [20]

Category	Method	Normal	Benign	UDH	ADH	FEA	DCIS	Invasive	Weighted F1
Human	Pathologists scores [20]	51.57 ± 11.70	52.15 ± 1.85	38.78 ± 10.22	37.89 ± 3.12	46.16 ± 19.64	68.26 ± 2.62	92.49 ± 2.14	56.36 ± 0.76
Supervised CNN Algorithms	CNN(10x) [20]	48.67 ± 1.71	44.33 ± 1.89	45.00 ± 4.97	24.00 ± 2.83	47.00 ± 4.32	53.33 ± 2.62	86.67 ± 2.64	50.85 ± 2.64
	CNN(40x) [20]	32.33 ± 4.64	39.00 ± 0.82	23.67 ± 1.70	18.00 ± 0.82	37.67 ± 2.87	47.33 ± 2.05	70.67 ± 0.47	39.41 ± 1.89
	Multi-scale CNN [20]	50.33 ± 0.94	44.33 ± 1.25	41.33 ± 2.49	31.67 ± 3.30	51.67 ± 3.09	57.33 ± 0.94	86.00 ± 1.41	52.83 ± 1.92
Supervised GNN Algorithms	CGC-Net [29]	30.83 ± 5.33	31.63 ± 4.66	17.33 ± 3.38	24.50 ± 5.24	58.97 ± 3.56	49.36 ± 3.41	75.30 ± 3.20	43.63 ± 0.51
	CG-GNN [20]	58.77 ± 6.82	40.87 ± 3.05	46.82 ± 1.95	39.99 ± 3.56	63.75 ± 10.48	53.81 ± 3.89	81.06 ± 3.33	55.96 ± 1.01
	TG-GNN [20]	63.59 ± 4.88	47.73 ± 2.87	39.41 ± 4.70	28.51 ± 4.29	72.15 ± 1.35	54.57 ± 2.23	82.21 ± 3.99	56.62 ± 1.35
	HACT-Net [20]	61.56 ± 2.15	47.49 ± 2.94	43.60 ± 1.86	40.42 ± 2.55	74.22 ± 1.41	66.44 ± 2.57	88.40 ± 0.19	61.53 ± 0.87
Supervised Baselines	CG	59.70 ± 4.54	46.64 ± 2.32	41.00 ± 2.41	36.26 ± 4.61	53.07 ± 1.85	54.69 ± 2.77	76.54 ± 1.86	53.29 ± 1.17
	PG	62.62 ± 4.81	44.86 ± 2.26	44.86 ± 1.91	25.92 ± 5.30	65.56 ± 1.96	47.99 ± 0.46	77.70 ± 2.50	53.72 ± 0.96
Self-Supervised Algorithms	ADGCL CG	50.61 ± 0.52	45.22 ± 1.67	35.41 ± 2.51	32.71 ± 3.10	58.24 ± 0.68	48.96 ± 2.50	61.13 ± 1.07	48.16 ± 0.16
	InfoGraph CG	51.26 ± 4.48	50.77 ± 0.84	37.53 ± 2.67	32.28 ± 1.86	56.33 ± 3.71	54.73 ± 1.07	74.46 ± 1.11	51.91 ± 0.69
	MV CG (Ours)	50.24 ± 1.82	50.43 ± 1.09	35.43 ± 3.32	35.43 ± 2.76	57.07 ± 3.10	57.59 ± 2.74	79.42 ± 3.71	**53.23 ± 0.18**
	ADGCL PG	50.32 ± 3.51	42.25 ± 2.30	32.69 ± 3.66	36.36 ± 3.74	57.97 ± 4.43	49.16 ± 3.15	77.15 ± 2.46	50.47 ± 1.92
	InfoGraph PG	56.84 ± 3.20	41.95 ± 1.12	36.75 ± 1.96	39.20 ± 1.12	53.56 ± 0.57	55.80 ± 3.97	79.34 ± 2.81	52.83 ± 1.60
	MV PG (Ours)	53.89 ± 3.08	42.12 ± 0.74	40.12 ± 2.91	42.23 ± 4.13	61.87 ± 1.18	59.23 ± 1.13	83.35 ± 1.09	**55.76 ± 0.97**

Table 2. BRACS ablation study for sampling strategies

Method	Normal	Benign	UDH	ADH	FEA	DCIS	Invasive	Weighted F1
No Sampling CG	49.19 ± 3.93	52.92 ± 0.29	38.83 ± 2.44	35.75 ± 1.22	58.79 ± 0.54	55.82 ± 2.42	76.92 ± 1.17	53.52 ± 0.68
Random Sampling CG	49.95 ± 4.63	52.02 ± 0.71	35.70 ± 2.96	33.99 ± 1.68	59.44 ± 3.18	58.76 ± 1.35	79.87 ± 1.34	**53.86 ± 0.92**
Similar Sampling CG	47.86 ± 4.78	50.82 ± 1.24	36.91 ± 3.17	36.62 ± 3.36	59.21 ± 4.82	57.33 ± 0.88	76.06 ± 5.09	53.07 ± 1.50
Dissimilar Sampling CG	48.30 ± 1.22	50.79 ± 1.47	33.30 ± 2.33	36.88 ± 3.58	56.83 ± 3.21	55.80 ± 1.13	80.14 ± 2.90	52.77 ± 0.79
Similar+Random Sampling CG	49.00 ± 3.77	50.42 ± 2.25	38.70 ± 2.52	34.10 ± 1.89	58.14 ± 3.06	54.71 ± 5.80	76.81 ± 3.11	52.63 ± 0.87
MV CG (Ours)	50.24 ± 1.82	50.43 ± 1.09	35.43 ± 3.32	35.43 ± 2.76	57.07 ± 3.10	57.59 ± 2.74	79.42 ± 3.71	53.23 ± 0.18
No Sampling PG	50.35 ± 3.13	42.21 ± 2.40	34.30 ± 0.68	36.22 ± 3.75	60.87 ± 3.02	55.92 ± 2.92	82.20 ± 0.93	52.91 ± 1.11
Random Sampling PG	48.79 ± 4.27	40.06 ± 0.99	35.66 ± 4.05	42.85 ± 3.14	61.97 ± 1.70	55.84 ± 3.16	80.21 ± 0.77	53.33 ± 1.25
Similar Sampling PG	50.00 ± 3.90	43.78 ± 0.63	37.61 ± 2.84	38.33 ± 2.76	58.30 ± 3.69	57.09 ± 3.88	83.12 ± 1.37	53.69 ± 1.51
Dissimilar Sampling PG	55.14 ± 2.19	40.24 ± 2.55	37.40 ± 2.45	39.14 ± 2.69	61.34 ± 2.42	58.25 ± 2.14	83.23 ± 1.55	54.66 ± 1.97
Similar+Random Sampling PG	49.50 ± 2.79	41.80 ± 3.02	37.59 ± 5.22	34.09 ± 1.92	60.09 ± 0.27	55.60 ± 0.83	80.34 ± 2.09	52.39 ± 1.14
MV PG (Ours)	53.89 ± 3.08	42.12 ± 0.74	40.12 ± 2.91	42.23 ± 4.13	61.87 ± 1.18	59.23 ± 1.13	83.35 ± 1.09	**55.76 ± 0.97**

labeled *Supervised Baselines*. Furthermore, we compare our results with other supervised algorithms on the dataset as an anchor, as indicated in the row labeled *Supervised Algorithms*. We also evaluate the representations generated by the SimCLR pretrained Resnet34 network, which provides node-level representations for our graph nodes. We apply mean pooling on the representations of individual patches extracted from the TRoI to obtain a graph-level representation. Using logistic regression, we achieve a cumulative weighted F1-score of 45.27% for this approach.

Table 1 presents results on the BRACS dataset on the 7-way classification downstream task. We report the mean and standard deviation of the weighted F1 score for each model on the test set by training them three times across different random seeds. Our method outperforms other self-supervised approaches across both cell graphs and patch graphs, achieving a weighted F1 score of 55.76±0.97% for Multiview Patchgraph (MV PG) and 53.23 ± 0.18% for Multiview Cellgraph

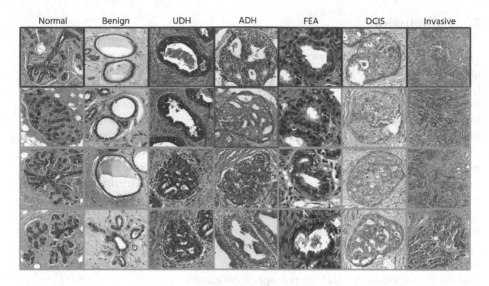

Fig. 2. Retrieving image patches from the BRACS testset based on *MV PG* graph representation of query images shown in blue boxes. The classes of the query images are mentioned in the top row. The closest representations according to Euclidean distance are placed below the query image. Among the retrieved images, the incorrect class is marked in red, and the correct class in green. Although the incorrectly retrieved image belongs to the DCIS class, we see a lot of visual similarities with the query image belonging to the ADH class

(MV CG). The superior performance of MV over its single-view counterpart Info-Graph highlights the benefits of utilizing multiple graph views for improved representations. We hypothesize ADGCL's performance deterioration compared to InfoGraph, and the proposed method is due to its edge-dropping augmentation, which is unsuitable for the downstream task, as it depends significantly on the topology of biological entities. Moreover, MV PG surpasses several supervised learning algorithms, including supervised baselines, CNN, Multi-Scale CNN, and CGC-Net [29], and comes close to pathologists' scores. We also present ablation results in Table 2, comparing different sampling strategies. *"Similar Sampling"* refers to picking cell graph nodes with a high dot product with the patch graph representation, while *"Dissimilar Sampling"* refers to the vice versa approach. Overall, our *"Random + Dissimilar"* node sampling strategy denoted by *MV PG/CG* outperforms other strategies for patch graphs and performs similarly for cell graphs. We hypothesize that maximizing mutual information with dissimilar nodes is akin to learning from hard samples, enabling the patch graph representation to incorporate diverse information (Table 2).

5   Conclusion

We introduce a self-supervised multiview GRL approach that utilizes different graph views generated from histology images to create comprehensive representations achieving superior performance compared to other self-supervised GRL methods and comes close to the pathologists' scores and other supervised learning algorithms. Our methodology provides an alternative to using graph augmentations for contrastive learning and helps in capturing different inductive biases introduced by different graph views into a single representation. While we demonstrate our model's efficacy on TRoIs, our algorithm is scalable to performing SSL on the WSI level, and we plan to leave it for future work. Future work also includes investigating the number and type of graph views that can be utilized to generate richer representations. Our highly versatile and modular framework allows users to leverage large unlabelled datasets and pre-train their GNN encoders by adding additional graph views to their existing input-output pipelines. This allows users to introduce prior knowledge without changing their desired input, encoder, and output specifications.

References

1. Ahmedt-Aristizabal, D., Armin, M.A., Denman, S., Fookes, C., Petersson, L.: A survey on graph-based deep learning for computational histopathology. Comput. Med. Imaging Graph. **95**, 102027 (2022)
2. Anklin, V., et al.: Learning whole-slide segmentation from inexact and incomplete labels using tissue graphs. In: Medical Image Computing and Computer Assisted Intervention-MICCAI 2021: 24th International Conference, Strasbourg, France, September 27-October 1, 2021, Proceedings, Part II 24, pp. 636–646. Springer (2021)
3. Aygüneş, B., Aksoy, S., Cinbiş, R.G., Kösemehmetoğlu, K., Önder, S., Üner, A.: Graph convolutional networks for region of interest classification in breast histopathology. In: Medical Imaging 2020: Digital Pathology, vol. 11320, pp. 134–141. SPIE (2020)
4. Biewald, L.: Experiment tracking with weights and biases (2020). https://www.wandb.com/, software available from wandb.com
5. Brancati, N., et al.: Bracs: A dataset for breast carcinoma subtyping in h&e histology images. Database 2022 (2022)
6. Chen, R.J., Lu, M.Y., Wang, J., Williamson, D.F., Rodig, S.J., Lindeman, N.I., Mahmood, F.: Pathomic fusion: an integrated framework for fusing histopathology and genomic features for cancer diagnosis and prognosis. IEEE Trans. Med. Imaging **41**(4), 757–770 (2020)
7. Chen, T., Kornblith, S., Norouzi, M., Hinton, G.: A simple framework for contrastive learning of visual representations. In: International Conference on Machine Learning, pp. 1597–1607. PMLR (2020)
8. Ciga, O., Xu, T., Martel, A.L.: Self supervised contrastive learning for digital histopathology. Mach. Learn. Appl. **7**, 100198 (2022)
9. Eldar, Y., Lindenbaum, M., Porat, M., Zeevi, Y.Y.: The farthest point strategy for progressive image sampling. IEEE Trans. Image Process. **6**(9), 1305–1315 (1997)

10. Fey, M., Lenssen, J.E.: Fast graph representation learning with pytorch geometric. arXiv preprint arXiv:1903.02428 (2019)
11. Gamper, J., et al.: Pannuke dataset extension, insights and baselines. arXiv preprint arXiv:2003.10778 (2020)
12. Gasteiger, J., Weißenberger, S., Günnemann, S.: Diffusion improves graph learning. Advances in neural information processing systems 32 (2019)
13. Graham, S., et al.: Hover-net: simultaneous segmentation and classification of nuclei in multi-tissue histology images. Med. Image Anal. **58**, 101563 (2019)
14. Hassani, K., Khasahmadi, A.H.: Contrastive multi-view representation learning on graphs. In: International Conference on Machine Learning, pp. 4116–4126. PMLR (2020)
15. He, K., Zhang, X., Ren, S., Sun, J.: Deep residual learning for image recognition. In: Proceedings of the IEEE Conference on Computer Vision and Pattern Recognition, pp. 770–778 (2016)
16. Lee, Y., Park, J.H., Oh, S., Shin, K., Sun, J., Jung, M., Lee, C., Kim, H., Chung, J.H., Moon, K.C., et al.: Derivation of prognostic contextual histopathological features from whole-slide images of tumours via graph deep learning. Nature Biomedical Engineering pp. 1–15 (2022)
17. Nowozin, S., Cseke, B., Tomioka, R.: f-gan: Training generative neural samplers using variational divergence minimization. Advances in neural information processing systems 29 (2016)
18. Ozen, Y., Aksoy, S., Kösemehmetoğlu, K., Önder, S., Üner, A.: Self-supervised learning with graph neural networks for region of interest retrieval in histopathology. In: 2020 25th International Conference on Pattern Recognition (ICPR), pp. 6329–6334. IEEE (2021)
19. Paszke, A., Gross, S., Massa, F., Lerer, A., Bradbury, J., Chanan, G., Killeen, T., Lin, Z., Gimelshein, N., Antiga, L., et al.: Pytorch: An imperative style, high-performance deep learning library. Advances in neural information processing systems 32 (2019)
20. Pati, P., Jaume, G., Foncubierta-Rodriguez, A., Feroce, F., Anniciello, A.M., Scognamiglio, G., Brancati, N., Fiche, M., Dubruc, E., Riccio, D., et al.: Hierarchical graph representations in digital pathology. Med. Image Anal. **75**, 102264 (2022)
21. Pina, O., Vilaplana, V.: Self-supervised graph representations of wsis. In: Geometric Deep Learning in Medical Image Analysis, pp. 107–117. PMLR (2022)
22. Srinidhi, C.L., Kim, S.W., Chen, F.D., Martel, A.L.: Self-supervised driven consistency training for annotation efficient histopathology image analysis. Med. Image Anal. **75**, 102256 (2022)
23. Sun, F.Y., Hoffmann, J., Verma, V., Tang, J.: Infograph: Unsupervised and semi-supervised graph-level representation learning via mutual information maximization. arXiv preprint arXiv:1908.01000 (2019)
24. Suresh, S., Li, P., Hao, C., Neville, J.: Adversarial graph augmentation to improve graph contrastive learning. Adv. Neural. Inf. Process. Syst. **34**, 15920–15933 (2021)
25. Vahadane, A., Peng, T., Sethi, A., Albarqouni, S., Wang, L., Baust, M., Steiger, K., Schlitter, A.M., Esposito, I., Navab, N.: Structure-preserving color normalization and sparse stain separation for histological images. IEEE Trans. Med. Imaging **35**(8), 1962–1971 (2016)
26. Wang, J., Chen, R.J., Lu, M.Y., Baras, A., Mahmood, F.: Weakly supervised prostate tma classification via graph convolutional networks. In: 2020 IEEE 17th International Symposium on Biomedical Imaging (ISBI), pp. 239–243. IEEE (2020)
27. Xu, K., Hu, W., Leskovec, J., Jegelka, S.: How powerful are graph neural networks? arXiv preprint arXiv:1810.00826 (2018)

28. You, Y., Chen, T., Sui, Y., Chen, T., Wang, Z., Shen, Y.: Graph contrastive learning with augmentations. Adv. Neural. Inf. Process. Syst. **33**, 5812–5823 (2020)
29. Zhou, Y., Graham, S., Alemi Koohbanani, N., Shaban, M., Heng, P.A., Rajpoot, N.: Cgc-net: Cell graph convolutional network for grading of colorectal cancer histology images. In: Proceedings of the IEEE/CVF International Conference on Computer Vision Workshops (2019)
30. Zhou, Z., Hu, Y., Zhang, Y., Chen, J., Cai, H.: Multiview deep graph infomax to achieve unsupervised graph embedding. IEEE Trans. Cybern. (2022)

Multi-level Graph Representations of Melanoma Whole Slide Images for Identifying Immune Subgroups

Lucy Godson[1]([✉]), Navid Alemi[1], Jérémie Nsengimana[5], Graham P. Cook[3], Emily L. Clarke[2,4], Darren Treanor[2,4], D. Timothy Bishop[3], Julia Newton-Bishop[3], and Derek Magee[1]

[1] School of Computing, University of Leeds, Woodhouse, Leeds LS2 9JT, UK
sclg@leeds.ac.uk
[2] Division of Pathology and Data Analytics, Leeds Institute of Cancer and Pathology, University of Leeds, Beckett Street, Leeds LS9 7TF, UK
[3] Leeds Institute of Medical Research at St James's, University of Leeds, Leeds LS2 9JT, UK
[4] Department of Histopathology, Leeds Teaching Hospitals Trust, Leeds LS2 9JT, UK
[5] Population Health Sciences Institute, Newcastle University, Newcastle upon Tyne NE1 7RU, UK

Abstract. Stratifying melanoma patients into immune subgroups is important for understanding patient outcomes and treatment options. Current weakly supervised classification methods often involve dividing digitised whole slide images into patches, which leads to the loss of important contextual diagnostic information. Here, we propose using graph attention neural networks, which utilise graph representations of whole slide images, to introduce context to classifications. In addition, we present a novel hierarchical graph approach, which leverages histopathological features from multiple resolutions to improve on state-of-the-art (SOTA) multiple instance learning (MIL) methods. We achieve a mean test area under the curve metric of 0.80 for classifying low and high immune melanoma subtypes, using multi-level and 20x patch graph representations of whole slide images, compared to 0.77 when using SOTA MIL methods. Our experimental results comprehensively show how our whole slide image graph representation is a valuable improvement on the MIL paradigm and could help to determine early-stage prognostic markers and stratify melanoma patients for effective treatments. Code is available at https://github.com/lucyOCg/graph_mil_project/.

Keywords: Computational pathology · graph neural networks · melanoma

1 Introduction

Melanoma is the most aggressive form of skin cancer [15], however, immunotherapy has revolutionised the treatment of patients [5]. Yet, the most effective

S.-A. Ahmadi and S. Pereira (Eds.): MICCAI 2023, LNCS 14373, pp. 85–96, 2024.
https://doi.org/10.1007/978-3-031-55088-1_8

and well tolerated drug, PD-1 blockade only benefits around 35% of patients [19]. Increasing our understanding of the interaction between immune and tumour cells and being able to identify disease subtypes is vital for stratifying patients into effective treatment groups and improving outcomes. Through consensus clustering of patient transcriptomes, previous studies have found distinct immunological subgroups within a population ascertained cohort (the Leeds Melanoma Cohort [LMC]), with differing clinical outcomes and potential treatment targets [16,18].

While tumour transcriptome data analysis is not routinely carried out for melanoma patients, haematoxylin and eosin (H&E) histopathology slides are well established in the diagnostic workflow of patients. Moreover, convolutional neural networks (CNNs), have been shown to identify molecular immune cell signatures from morphological patterns in digitised WSIs [22]. Analysis of these whole slide images (WSIs) can be challenging, as they are multi-resolution and multi-gigabyte, so current classification methods such as multiple instance learning (MIL) divide these images into patches, which are processed individually, but this leads to a loss of contextual information.

Here we propose a novel patch-based graph method, that exploits the intrinsic spatial positioning of all patches within a WSI and is also memory efficient, using patch-level feature embeddings from a CNN. In addition, we take inspiration from hierarchical cell-graphs [11,13,17,23], developing a multi-level approach, which exploits the inherent hierarchical relationships between features extracted from patches at different resolutions. We believe that the addition of graph features increases the contextual information learned by the models for classifying melanomas into immune subgroups and demonstrate how this leads to improved performance over state-of-the-art MIL methods for our novel application (Fig. 1).

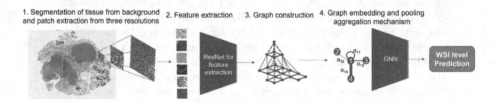

Fig. 1. Experimental framework for classifying the WSIs, using graph representations.

2   Related Works

WSIs are multi-resolution gigapixel images, which means they can be computationally and memory intensive to process, especially on GPUs. Moreover, most WSIs will only have slide-level labels, as pixel-wise labelling can be time consuming for a pathologist. MIL frameworks, where an image is treated as a bag with instances which inherit the bag or slide-level label [6], have been applied with high accuracy in computational pathology tasks for classifying WSIs [2,12,21]. In this method, a histopathology image can be subdivided into smaller patches, which can be further processed, using convolution neural networks (CNNs), to create feature representations. Following this, different pooling functions, such as maximum or mean operators [10], can be applied to the features to estimate the slide-level label classification.

While weakly supervised methods like MIL address problems such as the lack of annotation within WSIs and the aggregation of patch instances within an image, they also lead to a loss of contextual information. Each patch-instance or patch-feature is treated individually as it is passed through a network, with the focus being on the local visual and morphological patterns within the patch region. This means we lose global information about the tumour architecture, which can be important when evaluating the role of immune cells within the tumour microenvironment (TME), as they can differ depending on their spatial arrangement, locations and interactions with other cell types within the TME.

One way to resolve this, is through representing histopathological features in a graph structure and using a graph pooling mechanism. Recent works have shown how graph neural networks (GNN) can be a powerful tool for WSI analysis and subtyping [11,13,17,23]. However, these methods often require sampling techniques to select patches due to memory constraints and therefore can lose the overall WSI structure. Moreover, as early as 2011, [20] showed how multi-scale WSI analysis could be implemented for segmentation tasks and [17,23] demonstrated how multi-level hierarchical graphs can improve classification task performance. In addition, a study by [9], showed how GNNs using multi-scale WSI graph representations constructed with patch embeddings, could be used for grading and subtyping esophageal cancer and kidney cancer TCGA datasets. Here we build on this idea, but also look at how one-hot encodings for node and edge features can be used to generate novel patch-based graph representations, which have a global hierarchical "multi-level" structure.

3   Dataset

The dataset used for our work was from the Leeds Melanoma Cohort [14]. This is a population ascertained cohort, including 667 digitised WSIs of primary melanoma tumours. The labels for the images were delineated by clustering transcriptomes, based on the inferred abundance of 27 immune cell types [18]. The three subgroups are the "high immune" class (22.5%), which corresponds to greater inferred immune cell infiltration in the primary tumour and better associated patient survival outcomes, the "intermediate immune" class (39.1%), which corresponds to less inferred immune cell infiltrate in the primary tumour and the "low immune" class (38.4%), which has the least inferred immune cell infiltrate in the tumour and worst survival response of patients. We worked under the assumption that each group with a distinct immune genetic signature would have a distinct histological pattern. Initial experiments were carried out by training models using the three subgroups, but we also examined model performance when training and testing the models using only the "high" and "low immune" as these groups are more well defined compared to the "intermediate" subgroup which is highly heterogeneous. Furthermore the "high" and "low immune" subgroup are more informative for immunotherapy options.

Table 1. Summary data of the clinical and clinicohistological features of the three LMC immune subgroups found in the paper published by [18].

	Low	Intermediate	High
Number of tumours	256	261	150
Melanoma death (%)	36	27	21
Age at diagnosis (median, years)	58	58	60
Sex (% males)	43	45	49
Breslow thickness (median, mm)	2.45	2.29	2.00
Ulcerated (%)	30	34	35
AJCC stage (%)	Low	Intermediate	High
I	30	36	39
II	50	50	47
III	18	13	14
Unknown	2	2	1

All slides come from Formalin-Fixed Paraffin-Embedded (FFPE) blocks and were scanned in batches using a Leica Biosystems Aperio Digital Pathology Slide Scanner, at 0.25 micrometers-per-pixel (m.p.p.). The tumour transcriptomic data that was used to develop the immune subgroup labels was produced from the archived FFPE tumour blocks, using Illumina array DASL HT12.4 and normalised using standard methods as described in the study by [16]. In addition the clinical and prognostic features for the subgroups are shown in Table 1. Breslow thickness and microscopic ulceration (shown as "Ulcerated" in Table 1) are histologic features that have been shown to be a strong independent predictors of melanoma death [8]. Furthermore, while the LMC study aimed to include patients from diverse ethnic backgrounds, 99% of the participants were of Caucasian ethnicity. As a result of this, the predictive models developed in this study may exhibit a bias towards classifying patients with lighter skin and may not be easily generalised to other ethnicities and populations.

4   Methods

4.1   Segmentation and Feature Extraction

The H&E stained tissue in the WSI was segmented from the background using thresholding and morphological operations described by [12]. The segmented tissue was then split into 256 pixel × 256 pixel non-overlapping patches at three different resolutions (10x, 20x and 40x). A ResNet18, which had been pretrained by self-supervised learning on 57 multi-organ, multi-resolution histopathology datasets by [3], was used to extract 512-dimensional feature embeddings from the patches.

4.2   Graph Construction

We constructed global graph representations $G = (V,E)$ of the WSIs, where G is the graph, $V = \{1,...,n\}$ is the set of n vertices or nodes, which correspond to the n patch instances in a WSI. $E \subseteq V \times V$ is the set of edges, where $(i,j) \subseteq E$ is an edge that connects node i to node j. Node features for the graphs consisted of the 512-dimensional feature embeddings extracted from the patches. For experiments using single-resolution graph features, edge connections were defined between any vertical and horizontal neighbouring patches. To represent uniform edges between adjacent patches in single-resolution graph representations, edge features were set to 1. We also wanted to build what we term "multi-level" graph representations, where node features consisted of the extracted features from each of the three resolutions, and edges included those between adjacent patches within a single resolution and between patches from different resolutions. This involved defining edge connections between 10x patch nodes and the four spatially corresponding 20x patch nodes in the higher resolution level. Additionally, edge connections are defined between the 20x patch nodes to the four spatially corresponding 40x patch nodes to create hierarchical multi-resolution patch graphs. We also experimented with adding one-hot encodings to the node embeddings to distinguish between nodes of different resolutions (10x: [1,0,0], 20x: [0,1,0] and 40x: [0,0,1]). These encodings were concatenated to the node feature embeddings giving a 515-dimensional embedding. All nodes were connected in both directions, and self-loops were included. To distinguish between edges that connect patches from different resolutions and edges that connect adjacent patches within the same resolution, edge features were encoded using a one-hot encoding scheme (Fig. 2).

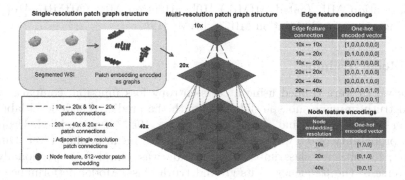

Fig. 2. Single and multi-resolution patch graph representation construction and edge feature encodings used for multi-resolution graphs. Self-loops are not visualised.

4.3 Model Architecture

When implementing the GNNs, we used GATv2 graph attention layers with one attention head [1]. Edge indices, node and edge features were passed through the GATv2 layers. A scoring function $e : \mathbb{R}^d \times \mathbb{R}^d \rightarrow \mathbb{R}$ was used to evaluate the importance of each edge (j,i) and the importance between the features of the neighbour node j to node i. The attention coefficients were formulated using:

$$e(\boldsymbol{h}_i, \boldsymbol{h}_j, \boldsymbol{e}_{ij}) = LeakyReLU((\boldsymbol{a} \odot \boldsymbol{W})^T[\boldsymbol{h}_i||\boldsymbol{h}_j||\boldsymbol{e}_{ij}]) \qquad (1)$$

Where $\boldsymbol{h}_i$ is a node, $\boldsymbol{h}_j$ is a neighbouring node and $\boldsymbol{e}_{ij}$ represents the edge features between them. $\boldsymbol{W}$ represents the weights matrix, $\boldsymbol{a}$ represents learnable attention weights, $\odot$ denotes element-wise multiplication and $||$ denotes vector concatenation. The attention scores are then scaled for all nodes within an image $j \in \mathcal{N}_i$, using a *softmax* function in the attention mechanism:

$$softmax_j(\boldsymbol{a}_{ij}) = \frac{exp(e(\boldsymbol{h}_i, \boldsymbol{h}_j, \boldsymbol{e}_{ij}))}{\sum_{k \in \mathcal{N}_i} exp(e(\boldsymbol{h}_i, \boldsymbol{h}_k, \boldsymbol{e}_{ik}))} \qquad (2)$$

where $\mathcal{N}_i$ represents the set of neighbours of node i and $\boldsymbol{a}_{ij}$ is the attention coefficient between node i and node j. $\boldsymbol{k}$ represents a node index from the set of all neighbors $\mathcal{N}_i$ of node i. The output of the GATv2 layer is represented by:

$$h'_i = \sigma \left(\sum_{j \in \mathcal{N}_i} e(\boldsymbol{h}_i, \boldsymbol{h}_i, \boldsymbol{e}_{ij}) \cdot (\mathbf{W} \odot \mathbf{a})h_j \right) \qquad (3)$$

where h'_i is the updated representation of node i after the GATv2 layer operation. σ represents the Exponential Linear Unit (*ELU*) activation function [4], which was followed by a global mean pooling operation. The number of GATv2 layers was varied to assess impact on performance. The outputs from each global mean pooling operation, when experimenting with multiple GATv2 layers, were then concatenated together and passed through two fully connected layers and a *softmax* activation function to give the classification output. We compared the GNNs, to four weakly supervised MIL methods, including max-pooling MIL, gated-attention MIL (gated AttMIL) [10], attention MIL (AttMIL) [10] and clustering-constrained attention MIL (CLAM) [12].

4.4 Model Training

The networks were trained using a cross-entropy loss function, comparing the ground truth immune subtype slide label with the predicted slide-level label. A learning rate of 0.0002 and weight decay of $1 \times 10e^{-5}$ were applied. Dropout with a probability of 0.5 was used after each global mean pooling layer. During training, to mitigate class imbalances, a slide was sampled proportionally to the inverse of the frequency of its ground truth class. Model performance was evaluated using a 10-fold Monte Carlo cross validation approach to calculate the mean area under the curve (AUC) with 95% confidence intervals (CI). For

each fold the data was split with 80% of the data being used for training data
and 10% being kept for both the test and validation datasets. When calculating
performance for the three immune subgroups, the AUC scores were calculated
for individual classes by binarising classifications, then averaging the AUCs for
the three classes. The models were trained for a minimum of 50 epochs, with
early stopping if the validation loss did not improve for 20 epochs continuously.

Graph models were implemented using PyTorch 1.13.1, PyTorch geometric
version 2.2.0 with CUDA 11.6, using one Nvidia V100 32GB GPU from the
JADE2 HPC facility. Segmentation, feature extraction and MIL models were
implemented using PyTorch version 1.7.1 with CUDA version 11.0, using one
Tesla V100 32G NVLink 2.0 GPU from the Bede N8 HPC facility.

5   Results

For classifying the three immune subtypes, the combination of graph and visual
features led to improved performance, when utilising single resolution and multi-
resolution graphs with one-hot encoded node embeddings (Table 2). The best
performing model achieved a mean test AUC of 0.63 (95% CI 0.61 to 0.65),
using 10x single resolution WSI graph representations. This may be the case due
to lower resolution 10x patches containing a balance of cellular and structural
detail. Furthermore, as melanoma is a highly heterogeneous tumour, a greater
number of nodes can contribute to increased node heterogeneity, which leads to
noise that contributes to classifications errors. Moreover, this heterogeneity is
increased when including the "intermediate" subtype in the classification task,
as this subgroup is less defined than the "high" and "low immune" subtypes. Consequently, lower resolution 10x graph representations will contain fewer nodes
and therefore fewer noisy or heterogeneous nodes, so will generate better classification performance for the three subtype task.

Table 2. Mean test AUC for classification of three immune subtypes with 10 fold cross
validation with 95% CI (mean ± 95% CI), for different MIL models compared to our
proposed GNN models using single and multi-resolution graph representations. The
GNN models here are implemented with 3 GATv2 layers. "One-hot node" indicates
multi-resolution graphs with one-hot encodings added to the node patch embeddings.

Resolution	Max-pooling MIL	Gated AttMIL	AttMIL	CLAM	Proposed GNN
10x	0.57±0.022	0.60±0.026	0.61±0.039	0.60±0.028	**0.63±0.023**
20x	0.55±0.026	0.57±0.035	0.59±0.027	0.57±0.034	0.62±0.021
40x	0.48±0.023	0.55±0.036	0.54±0.023	0.55±0.035	0.61±0.027
10x+20x+40x	-	-	-	-	0.61±0.025
One-hot node	-	-	-	-	0.62±0.017

When examining model performance for classifying "high" and "low" immune
subtypes, we found that models trained with 10x and 20x single resolution graphs

and multi-resolution graphs outperformed current SOTA MIL models (Table 3). Moreover, we found that the best performing GNN models both achieved a mean AUC of 0.80 and were trained using 20x patch WSI graph representations, or one-hot encoded multi-resolution WSI graph representations. We believe this may be the case, as the presence of immune cells in the TME is important for classification of the "high immune" class, therefore, cellular and spatial detail is important. As graph representations provide spatial context and higher resolution patches provide cellular detail, a trade-off between enough cellular detail and a reduction in noisy uninformative patch node embeddings, is required. Hence, in Table 4 for single-resolution graphs, it appears that 20x patch graphs generate the best performance as they provide more cellular detail than 10x patches for the binary task, without being adversely affected by the noise seen with the 40x patch graph performance. Conversely, the one-hot node graph representations, while having the disadvantage of more noisy and heterogeneous nodes, have greater structural information from the edge and node one-hot encodings, which may be the reason that they generate the joint highest mean AUC when used as an input (0.80, [Table 4]).

Table 3. Mean test AUC for high v.s. low classification with 10 fold cross validation with 95% CI (mean ± 95% CI), for different MIL models compared to our proposed GNN models. Models were tested at different using single and multi-resolution graph representations. "one-hot node" indicates multi-resolution graphs with one-hot encodings added to the node patch embeddings. The 10x and 20x use three GATv2 layers, the 10x+20x+40x use two GATv2 layers and the GNNs with "one-hot node" and 40x embeddings use four GATv2 layers.

Resolution	Max-pooling MIL	Gated AttMIL	AttMIL	CLAM	Proposed GNN
10x	0.51±0.069	0.77±0.042	0.75±0.043	0.73±0.055	0.78±0.033
20x	0.57±0.090	0.75±0.072	0.74±0.048	0.75±0.053	**0.80±0.052**
40x	0.48±0.049	0.70±0.048	0.76±0.029	0.67±0.073	0.77±0.056
10x+20x+40x	-	-	-	-	0.78±0.042
One-hot node	-	-	-	-	**0.80±0.048**

Moreover, we tested how increasing the number of GATv2 layers affected model performance and how using only 10x and 20x patches to generate "multi-level" graphs affected performance (Table 4). Overall, We found that three GATv2 layers produced the best performance for the single and 10x-20x resolution GNN models. However, when implementing four GATv2 layers, the models trained with multi-resolution one-hot encoded node embeddings achieved the highest mean test AUC score of 0.80 (95% CI 0.75 to 0.85).

Table 4. Mean test AUC for high v.s. low classification with 10 fold cross validation with 95% CI (mean ± 95% CI), for GNNs when increasing the number of GATv2 layers in the network. "One-hot node" indicates multi-resolution graphs with one-hot encodings added to the node patch embeddings.

Resolution	1 GATv2 layer	2 GATv2 layers	3 GATv2 layers	4 GATv2 layers
10x	0.76±0.050	0.76±0.041	0.78±0.033	0.78±0.038
20x	0.73±0.061	0.78±0.045	**0.80±0.052**	0.78±0.068
40x	0.68±0.061	0.75±0.050	0.75±0.084	0.77±0.056
10x+20x	0.73±0.056	0.76±0.042	0.79±0.033	0.77±0.044
10x+20x+40x	**0.77±0.091**	**0.78±0.042**	0.77±0.088	0.77±0.088
One-hot node	0.73±0.065	0.75±0.077	0.78±0.076	**0.80±0.048**

6   Discussion and Conclusion

Recent studies [16, 18] have shown that melanoma patients can be stratified into subgroups, with added prognostic value compared to AJCC melanoma staging systems [8]. However, these studies are carried out using transcriptomic data, which can be expensive and time consuming to analyse. In this paper we show how GNNs with graph representations of digitised H&E stained slides can be used to classify patients into immune subgroups. Moreover, while performance does not increase beyond a mean test AUC of 0.63, for classifying the three immune subtypes, we show that GNN models lead to improved performance over SOTA MIL models. In order to improve performance further, we will need to decipher tumour heterogeneity and complexity within the "intermediate" subgroup. To tackle this problem, we may need to look at further dividing this subgroup, as a previous study by [16] found two distinct subgroups which overlap with this "intermediate" group, or look at different techniques to learn more discriminant representations of the images.

For the task of classifying "high immune" and "low immune" subtypes we show that 20x graph representations and our novel one-hot node encoded multi-level graph representations generate the superior performance, with a mean test AUC performance of 0.80. In agreement with findings by [23], our study also demonstrates that increasing the number of GATv2 message passing layers in models appears to enhance information transfer through the network, leading to increased performance when using one-hot encoded node embeddings (Table 4). We also see this trend with 40x resolution graphs, suggesting these larger graph structures, containing 40x node embeddings, benefit more from increased message passing layers for information transfer.

One potential limitation of our work, which may be contributing to weakened performance of the multi-level graph representation models, is the graph mean pooling mechanism. Here, all node embeddings are averaged together, equally contributing to the final slide-level prediction. Due to the highly heterogenous nature of melanoma, there are likely to be uninformative regions within the images, especially as the ground truth labels that we are predicting are derived from a small 0.6-mm needle biopsy area. These noisy instances are likely to be more frequent in the multi-level graph representations due to the increased number of patch inputs from all three resolutions, leading to more misclassifications and a reduction in multi-level graph model performance. To address this concern, we propose the incorporation of a learned attention mechanism, which can emphasise the contributions of the most informative node embeddings and enhance both the performance and interpretability of the model.

Recent studies have also shown how elements of both graph and vision transformer models can be combined to classify WSIs, retaining positional encodings with detailed patch level features. In 2022, [24] developed a GPT model, which combined GNN layers and Transformer layers to classify lung cancer subtypes, outperforming other SOTA MIL and graph models. In addition, [7] developed a model, which utilised two independent "Efficient Graph-based Transformer" branches, which processed both low-resolution and high-resolution patch embeddings for cancer subtyping and metastasis detection. We believe that implementing transformer layers in our own GNNs, may be another way of improving model performance, by introducing self-attention and positional encodings that may help reduce errors caused by non-informative patch embeddings.

Overall, we present a comprehensive study highlighting the superiority of graph-based methods over MIL methods for the novel task of classifying melanoma WSIs into immune subgroups. These findings strongly suggest that graph-based techniques could be applied to a wide variety of other problems where MIL is regarded as the gold standard. Furthermore, we showcase the clinical utility of graph-based methods in stratifying patients into prognostic immune groups, suggesting these methods are superior when modelling spatial relationships within the TME.

Acknowledgements. This work was supported by the Engineering and Physical Sciences Research Council (EPSRC) [EP/S024336/1]; Cancer Research UK [C588/A19167, C8216/A6129, and C588/A10721 and NIH CA83115]; and the Medical Research Council [MR/S001530/1]. This work also made use of the WSIs which were digitised by the National Pathology Imaging Co-operative, NPIC (Project no. 104687) which is supported by a £50m investment from the Data to Early Diagnosis and Precision Medicine strand of the government's Industrial Strategy Challenge Fund, managed and delivered by UK Research and Innovation (UKRI). This work made use of the facilities of the N8 Centre of Excellence in Computationally Intensive Research (N8 CIR) provided and funded by the N8 research partnership and EPSRC [EP/T022167/1] made use of time on Tier 2 HPC facility JADE2, funded by EPSRC (EP/T022205/1). We would like to also thank the LMC patients for their involvement and generosity in providing data for this study.

References

1. Brody, S., Alon, U., Yahav, E.: How Attentive are Graph Attention Networks? January 2022. http://arxiv.org/abs/2105.14491, arXiv:2105.14491 [cs] version: 3
2. Campanella, G., et al.: Clinical-grade computational pathology using weakly supervised deep learning on whole slide images. Nature Med. **25**(8), 1301–1309 (2019). https://doi.org/10.1038/s41591-019-0508-1. https://www.nature.com/articles/s41591-019-0508-1, number: 8 Publisher: Nature Publishing Group
3. Ciga, O., Xu, T., Martel, A.L.: Self supervised contrastive learning for digital histopathology. arXiv:2011.13971 [cs, eess], September 2021. http://arxiv.org/abs/2011.13971, arXiv: 2011.13971
4. Clevert, D.A., Unterthiner, T., Hochreiter, S.: Fast and Accurate Deep Network Learning by Exponential Linear Units (ELUs), February 2016. https://doi.org/10.48550/arXiv.1511.07289, http://arxiv.org/abs/1511.07289. arXiv:1511.07289 [cs] version: 5
5. Curti, B.D., Faries, M.B.: Recent Advances in the Treatment of Melanoma. New England J. Med., June 2021. https://doi.org/10.1056/NEJMra2034861. https://www.nejm.org/doi/10.1056/NEJMra2034861, publisher: Massachusetts Medical Society
6. Dietterich, T.G., Lathrop, R.H., Lozano-Pérez, T.: Solving the multiple instance problem with axis-parallel rectangles. Artif. Intell. **89**(1–2), 31–71 (1997). https://doi.org/10.1016/S0004-3702(96)00034-3. https://linkinghub.elsevier.com/retrieve/pii/S0004370296000343
7. Ding, S., Li, J., Wang, J., Ying, S., Shi, J.: Multi-scale Efficient Graph-Transformer for Whole Slide Image Classification, May 2023. https://doi.org/10.48550/arXiv.2305.15773, http://arxiv.org/abs/2305.15773, arXiv:2305.15773 [cs]
8. Gershenwald, J.E., Scolyer, R.A.: Melanoma Staging: American Joint Committee on Cancer (AJCC) 8th Edition and Beyond. Annal. Surg. Oncol. **25**(8), 2105–2110 (2018). https://doi.org/10.1245/s10434-018-6513-7
9. Hou, W., Yu, L., Lin, C., Huang, H., Yu, R., Qin, J., Wang, L.: H^2-MIL: exploring hierarchical representation with heterogeneous multiple instance learning for whole slide image analysis. In: Proceedings of the AAAI Conference on Artificial Intelligence 36(1), 933–941 (Jun 2022). https://doi.org/10.1609/aaai.v36i1.19976, https://ojs.aaai.org/index.php/AAAI/article/view/19976, number: 1
10. Ilse, M., Tomczak, J.M., Welling, M.: Attention-based Deep Multiple Instance Learning. arXiv:1802.04712 [cs, stat], June 2018. http://arxiv.org/abs/1802.04712, arXiv: 1802.04712
11. Lee, Y., Park, J.H., Oh, S., Shin, K., Sun, J., Jung, M., Lee, C., Kim, H., Chung, J.H., Moon, K.C., Kwon, S.: Derivation of prognostic contextual histopathological features from whole-slide images of tumours via graph deep learning. Nature Biomedical Engineering, August 2022. https://doi.org/10.1038/s41551-022-00923-0, https://www.nature.com/articles/s41551-022-00923-0
12. Lu, M.Y., et al.: Deep Learning-based Computational Pathology Predicts Origins for Cancers of Unknown Primary. Nature **594**(7861), 106–110 (2021). https://doi.org/10.1038/s41586-021-03512-4, http://arxiv.org/abs/2006.13932, arXiv:2006.13932 [cs, q-bio]
13. Lu, W., Toss, M., Rakha, E., Rajpoot, N., Minhas, F.: SlideGraph+: Whole Slide Image Level Graphs to Predict HER2Status in Breast Cancer, October 2021. http://arxiv.org/abs/2110.06042, arXiv:2110.06042 [cs]

14. Newton-Bishop, J.A., et al.: Serum 25-hydroxyvitamin D3 levels are associated with breslow thickness at presentation and survival from melanoma. J. Clin. Oncol. **27**(32), 5439–5444 (2009). https://doi.org/10.1200/JCO.2009.22.1135. https://ascopubs.org/doi/10.1200/JCO.2009.22.1135, publisher: Wolters Kluwer
15. NHS: Melanoma skin cancer, October 2017. https://www.nhs.uk/conditions/melanoma-skin-cancer/, section: conditions
16. Nsengimana, J., et al.: Beta-Catenin-mediated immune evasion pathway frequently operates in primary cutaneous melanomas, May 2018. https://doi.org/10.1172/JCI95351, https://www.jci.org/articles/view/95351/pdf, publisher: American Society for Clinical Investigation
17. Pati, P., et al.: Hierarchical graph representations in digital pathology. Med. Image Anal. **75**, 102264 (2022). https://doi.org/10.1016/j.media.2021.102264, https://www.sciencedirect.com/science/article/pii/S1361841521003091
18. Poźniak, J., et al.: Genetic and Environmental Determinants of Immune Response to Cutaneous Melanoma. Cancer Res. **79**(10), 2684–2696 (2019). https://doi.org/10.1158/0008-5472.CAN-18-2864. http://cancerres.aacrjournals.org/lookup/doi/10.1158/0008-5472.CAN-18-2864
19. Robert, C., et al.: Improved overall survival in melanoma with combined dabrafenib and trametinib. N. Engl. J. Med. **372**(1), 30–39 (2015). https://doi.org/10.1056/NEJMoa1412690
20. Roullier, V., Lézoray, O., Ta, V.T., Elmoataz, A.: Multi-resolution graph-based analysis of histopathological whole slide images: application to mitotic cell extraction and visualization. Computerized Medical Imaging and Graphics: The Official Journal of the Computerized Medical Imaging Society **35**(7–8), 603–615 (2011). https://doi.org/10.1016/j.compmedimag.2011.02.005
21. Schirris, Y., Gavves, E., Nederlof, I., Horlings, H.M., Teuwen, J.: DeepSMILE: Self-supervised heterogeneity-aware multiple instance learning for DNA damage response defect classification directly from H&E whole-slide images, July 2021. http://arxiv.org/abs/2107.09405, arXiv:2107.09405 [cs, eess]
22. Schmauch, B., et al.: A deep learning model to predict RNA-Seq expression of tumours from whole slide images. Nature Commun. **11**(1), 3877 (2020). https://doi.org/10.1038/s41467-020-17678-4. https://www.nature.com/articles/s41467-020-17678-4, number: 1 Publisher: Nature Publishing Group
23. Sims, J., Grabsch, H.I., Magee, D.: Using Hierarchically Connected Nodes and Multiple GNN Message Passing Steps to Increase the Contextual Information in Cell-Graph Classification. In: Manfredi, L., et al. (eds.) Imaging Systems for GI Endoscopy, and Graphs in Biomedical Image Analysis. pp. 99–107. Lecture Notes in Computer Science, Springer Nature Switzerland, Cham (2022). https://doi.org/10.1007/978-3-031-21083-9_10
24. Zheng, Y., et al.: A graph-transformer for whole slide image classification (May 2022), http://arxiv.org/abs/2205.09671, arXiv:2205.09671 [cs]

Heterogeneous Graphs Model Spatial Relationship Between Biological Entities for Breast Cancer Diagnosis

Akhila Krishna[✉], Ravi Kant Gupta, Nikhil Cherian Kurian, Pranav Jeevan, and Amit Sethi

Indian Institute of Technology Bombay, Mumbai, India
akhilakrishnak2000@gmail.com
https://www.iitb.ac.in/

Abstract. The heterogeneity of breast cancer presents considerable challenges for its early detection, prognosis, and treatment selection. Convolutional neural networks often neglect the spatial relationships within histopathological images, which can limit their accuracy. Graph neural networks (GNNs) offer a promising solution by coding the spatial relationships within images. Prior studies have investigated the modeling of histopathological images as cell and tissue graphs, but they have not fully tapped into the potential of extracting interrelationships between these biological entities. In this paper, we present a novel approach using a heterogeneous GNN that captures the spatial and hierarchical relations between cell and tissue graphs to enhance the extraction of useful information from histopathological images. We also compare the performance of a cross-attention-based network and a transformer architecture for modeling the intricate relationships within tissue and cell graphs. Our model demonstrates superior efficiency in terms of parameter count and achieves higher accuracy compared to the transformer-based state-of-the-art approach on three publicly available breast cancer datasets – BRIGHT, BreakHis, and BACH.

Keywords: Graph · Heterogeneous · Histology · Transformer

1 Introduction

Breast cancer is the most common cancer among women globally, and it continues to pose significant challenges for early diagnosis, prognosis, and treatment decisions, given its diverse molecular and clinical subtypes [25]. To address these challenges, recent advancements in machine learning techniques have paved the way for improved accuracy and personalized treatment strategies. However, most convolutional neural networks (CNNs) overlook the spatial relationships within histopathological images, treating them as regular grids of pixels [4,11,13,16]. To overcome this limitation, graph neural networks (GNNs) have emerged as a promising alternative for classifying breast cancer.

GNNs are designed to handle complex graph structures, making them well-suited for tasks that involve analyzing relationships between entities [15,22,23].

© The Author(s), under exclusive license to Springer Nature Switzerland AG 2024
S.-A. Ahmadi and S. Pereira (Eds.): MICCAI 2023, LNCS 14373, pp. 97–106, 2024.
https://doi.org/10.1007/978-3-031-55088-1_9

By representing histopathological images as graphs, with image regions and structures as nodes and their spatial relationships as edges, GNNs can capture the inherent spatial context within the images [2,8,17,20,24]. This allows them to extract valuable information and patterns that may be missed by other machine-learning methods.

In this paper, we propose the use of heterogeneous graph convolutions between the cell and the tissue graphs to enhance the extraction of spatial relationships within histology images. This approach allows for the incorporation of diverse, multi-scale, and comprehensive features and spatial relationships by modeling both cell and tissue structures as well as their hierarchical relationships. Specifically, we introduce three features that have not been previously used in GNNs for histopathology images: (1) heterogeneous graph convolutions along with a transformer which outperforms [12] on BRIGHT [6], (2) an adaptive weighted aggregation technique with heterogeneous convolutions that outperforms [12] and is more efficient in terms of the number of parameters, and (3) the cross-attention modules of CrossVit [7] along with heterogeneous graph convolutions to extract the spatial relationship between cell and tissue graphs. We also analyzed different methods of k-nearest neighbor (kNN) graph building for cell and tissue graphs and found that edges based on node feature similarities performed better than other methods such as spatial closeness [18,19] or dynamically learnable layers for edge creation [12]. Extensive experiments conducted on three publicly available breast cancer histology datasets – BRIGHT [6], BACH [3], and BreakHis [5] – demonstrate the gains of our method.

2   Related Work

The first work on entity-level analysis for histology was [24] in which a cell graph was created and a graph convolutional network was used for processing the graph. Other works, such as [12,18,19], have also focused on the entity-level analysis of histology images by constructing cell graphs and tissue graphs and by constructing more than two levels of subgraphs to capture the spatial relationship in the image. In [18], the authors introduced an LSTM-based feature aggregation technique to capture long-range dependencies within the graphs. By utilizing the LSTM, they aimed to capture the temporal dependencies between cells and tissues, enhancing the understanding of their relationships. In contrast, [12] took a different approach by constructing more than two levels of subgraphs and using a spatial hierarchical graph neural network. This network aimed to capture both long-range dependencies and the relationships between the cell and tissue graphs. To achieve this, the authors employed a transformer-based feature aggregation technique, leveraging the transformer's ability to capture complex patterns and dependencies in the data. However, it was observed that both of these approaches fell short of fully extracting the intricate relationship between tissue and cell graphs. Hence, there is a scope to explore alternative methods or combinations of techniques to better understand and leverage the spatial relationships within histology images for improved analysis and classification of breast cancer and other diseases.

3 Methodology

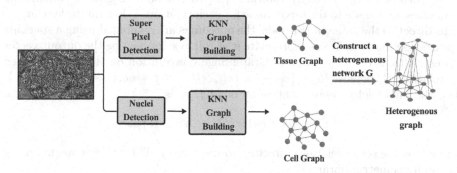

Fig. 1. Schematic of Graph Formation network. The image is fed into a superpixel detection network and a Hovernet-based nuclei detector to segment the image into superpixels and to identify the nuclei. Using this data, tissue and cell graphs are constructed with a k-nearest neighbor graph builder. These graphs are unified into a heterogenous graph by linking cell nodes to tissue nodes when the cell nuclei are part of the associated tissue region

3.1 Tissue and Cell Graph Extraction from Histology Images

The interrelationship between cells and tissue is investigated through the construction of cell and tissue graphs (Fig. 1). The goal is to perform multi-level feature analysis by extracting relevant features from these entities. To construct the cell graph, a pre-trained model called HoverNet [9] is used for nuclei detection. The feature representation for each nucleus is extracted by processing patches around the nuclei with a ResNet34 encoder [11,12]. Similarly, the tissue entities are identified using segmentation masks, which are obtained by applying the Simple Linear Iterative Clustering (SLIC) algorithm for superpixel segmentation [1,12]. Once the tissue entities are identified, their feature representations are extracted in the same manner as the cell entities. We then utilize kNN to get the k most similar nodes of all nodes based on the distance of node feature representations for the formation of edges between cell entities and tissue entities. In our experiment, we used $k = 5$ for all models and datasets. The edges between cells and tissues are formed by using the spatial position of cells and tissues. We treat a cell node C_i and a tissue node T_i as connected if $C_i \in T_i$.

3.2 Heterogeneous Graph Convolution

The interaction between cell and tissue graphs has to be captured effectively for better analysis which is done using heterogeneous graph convolutions. We define a heterogenous graph(H) as a union of cell-to-cell, tissue-to-tissue, and cell-to-tissue relations (which are defined by edges) along with their node features.

$$H = \{C, T, E_{cell->cell}, E_{tissue->tissue}, E_{cell->tissue}\} \tag{1}$$

where C and T are the features of the node in the cell graph and the tissue graph and $E_{A->B}$ is the list of edges between the elements in sets A and B.

We utilize Graph Sage convolutions [10] to transmit messages individually from the source node to the target node for each relation. When multiple relationships direct to the same target node, the outcomes are combined using a specified aggregation method. We can formulate it as follows: let z_{jR_i} be the output vector representations on node j due to Graph Sage convolution on the nodes defined by relation R_i, where $R_i \in \{cell-> cell, cell-> tissue, tissue-> tissue\}$, then the final vector representation on node j, z_j is

$$z_j = Aggregator(\{z_{jR_u}, \forall u \in U\}) \qquad (2)$$

where U is the set of relations directing to the node j. This is implemented using PyTorch geometric library.

3.3   Adaptive Weighted Aggregation for Multi-level Feature Fusion

Low-level and high-level features are important in classification tasks. So we use skip connections for its extraction and adaptive weighted aggregation layer for feature fusion. Adaptive weighted aggregation layer f is represented as follows:

$$f(w_1, w_2, ...w_n, F_1, F_2....F_n) = w_1F_1 + w_2F_2 + w_3F_3 + ...w_nF_n, \qquad (3)$$

where w_i is trainable weights of dimension 1×1 and F_i is feature vector of dimension 256×1 for $i \in 1, 2, 3, ...n$.

3.4   Cross-Attention Feature Fusion and Attentive Interaction Using Transformers

The transformer encoder of [21] which uses self-attention is used for extracting the long-range dependencies between the nodes in the graph. Formally, let us say the transformer encoder is T and the input to T is $Concat(N_c, N_t)$ where N_c and N_t are node features of cell and tissue graph respectively. The output from the transformer encoder goes to MLP layers to get the final output. N_c and N_t are of dimension $C \times 1 \times 256$ and $C \times 1 \times 256$ respectively where C is the number of nodes in the cell graph (we extrapolate the tissue graph with zeros for expanding the dimension of it to the same dimension as that of cell graph).

Cross-attention modules of [7] are used for extracting the interaction between cell graph and tissue graph which then is then fed into multi-layer perceptron layers to get the required output (Fig. 2).

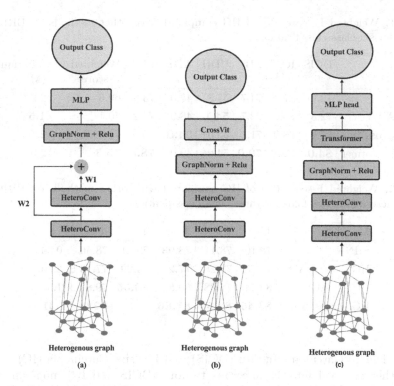

Fig. 2. Proposed architectural variants of Heterogeneous Graph Neural Network (HG): (a) HG with adaptive weighted aggregation of multi-level features, (b) HG with CrossVit for hierarchical feature fusion, and (c) HG with transformer for attentive interaction.

4   Experiments and Results

4.1   Clinical Datasets and Evaluation Methods

For assessing the utility of graph-based deep learning methods, breast cancer histology datasets form an ideal test bed due to the diversity of its subtypes. We used three datasets in this work. The BRIGHT dataset is a breast cancer subtyping dataset released as a part of the Breast tumor Image classification on Gigapixel Histopathological images challenge [6]. The breast cancer histopathology image datasets (BACH) [3] has four disease states – normal, benign, ductal carcinoma in-situ (DCIS), and invasive ductal carcinoma (IDC). The BreakHis [5] has malignant and non-malignant classes.

BRIGHT Dataset. The BRIGHT dataset contains 4025 hematoxylin & Eosin (H&E) stained breast cancer histology images scanned at 0.25 mn/pixel resolution. It is classified into 6 classes: Pathological Benign (PB), Usual Ductal Hyperplasia (UDH), Flat Epithelial Atypia (FEA), Atypical Ductal Hyperplasia

Table 1. Weighted F-score (%) of HG compared with other methods on BRIGHT dataset for six-class classification.

Model	DCIS	IC	PB	UDH	ADH	FEA	Weighted F-score	#Param (M)
SHNN	72.9	87.3	74.7	51.8	45.0	73.9	69.6	2.2
HG + AWA	72.8	88.0	72.7	**54.1**	48.9	74.2	70.1	**1.5**
HG + CrossVit	77.9	88.0	71.1	53.0	**54.0**	73.0	71.3	2.5
HG+transformer	**84.0**	**92.0**	**79.0**	54.0	49.0	**78.5**	**75.3**	2.0

Table 2. Weighted F-score (%) of HG compared with other methods on BRIGHT dataset across four test folds for three-class classification.

Model	1	2	3	4	$\mu \pm \sigma$
SHNN	78.10	78.11	78.93	79.21	78.56 ± 0.56
HG + CrossVit	78.75	78.28	77.2	74.29	77.13 ± 2.00
HG + AWA	80.24	77.98	80.82	**80.55**	79.89 ± 1.30
HG+transformer	**83.45**	**82.59**	**81.86**	80.36	**82.06±1.31**

(ADH), Ductal Carcinoma In Situ (DCIS) and Invasive Carcinoma (IC). These are further grouped into 3: cancerous tumors (DCIS and IC), non-cancerous tumors (PB and UDH) and pre-cancerous tumors (FEA and ADH) [6].

BACH Dataset. The BACH dataset contains 400 hematoxylin and eosin (H&E) stained breast cancer histology images with pixel scale $0.42\,\mu m \times 0.42\,\mu m$. It is classified into 4 classes: normal, benign, in situ carcinoma, and invasive carcinoma [3].

BreakHis Dataset. The BreakHis dataset contains 9109 microscopic images of breast cancer at 400X magnifying factor. It contains 2480 benign and 5429 malignant tissue images [5].

4.2   Experimental Setup

All the experiments are implemented in Pytorch and using Pytorch-geometric and histocartography library [14]. The F-score was used as the evaluation metric. We proposed three models and compared each other and also compared with the state-of-the-art model [12]. Hovernet was used for nuclei detection in all models for all datasets. The feature vector of length 512 for each nucleus was extracted from a patch size of 72, 72, and 48 around each nuclei using Resnet 34 model [11] for the BRIGHT, BACH, and BreakHis respectively. We also compared 3 methods of edge formation of graphs: dynamic structural learning [12], kNN algorithm on the distance between cell entities and the tissue entities, and kNN algorithm

Table 3. Weighted F-score (%) of HG compared with other methods on BRIGHT dataset across four test folds for three-class classification.

Model	Cancerous	Non-cancerous	Pre-cancerous
SHNN	85.21 ± 2.07	78.08 ± 0.81	72.11 ± 0.82
HG + CrossVit	83.93 ± 1.12	76.29 ± 3.08	70.38 ± 2.35
HG + AWA	87.08 ± 1.49	78.09 ± 2.64	73.75 ± 1.36
HG+transformer	**88.78±1.96**	**80.61±1.33**	**76.08 ±1.36**

Table 4. Weighted F-score (%) of HG compared with other methods on the BACH dataset.

Model	Normal	Benign	InSitu	Invasive	Weighted F-score
SHNN	**90.00**	80.00	85.10	84.21	84.83
HG+transformer	88.88	**86.48**	**85.20**	**90.10**	**86.62**

on node feature representation of cell entities and tissue entities. The batch size was set to 32. A learning rate of $1e^{-4}$ and Adam optimizer with a weight decay of $5e^{-4}$ were used while training the models. We have evaluated the model on 3 datasets. On the BRIGHT dataset, the method is evaluated across four test folds for three class classifications. Due to less data, we haven't evaluated the model across 4 test folds on BACH and BreakHis datasets.

4.3   Comparison with the State-of-the-Art Methods

The results for six-class classification and three-class classification on the BRIGHT dataset are listed in Tables 1 and 2 and 3 respectively. The four-class classification on the BACH dataset is shown in Table 4 and the 2-class classification on the BreakHis dataset is shown in Table 5. It can be observed that our heterogenous graph convolution and transformer-based network outperforms the SOTA by a considerable margin for all 6 classes of the BRIGHT dataset. The HG-Transformer network outperforms all the other models for DCIS, IC, PB, and FEA classes while HG-AWA and HG-CrossVit perform better for UDH and ADH respectively. With 1.5 million parameters, HG-AWA stands out as the most efficient model outperforming the SOTA. We have observed that even on BACH and BreakHis dataset, the proposed model is outperforming the graph based SOTA method.

4.4   Ablation Studies

We performed studies on different methods of graph formation and removing and adding different layers of the model. The results are listed in Table 6. The spatial hierarchical neural network (SHNN) of [12] is Graph Sage convolutions in homogeneous graphs with a transformer. It can be observed that the addition of a transformer to Graph Sage convolutions in the SHNN led to an increase in

Table 5. Weighted F-score (%) of HG compared with other methods on BreakHis dataset.

Model	Malignant	Benign	Weighted F-score
SHNN	89.16	78.17	85.46
HG+transformer	**95.62**	**91.70**	**94.30**

Table 6. Weighted F-score (%) of different graph-based methods compared on BRIGHT dataset.

Model	DCIS	IC	PB	UDH	ADH	FEA	Weighted F-score
Graph Sage Conv	72.1	84.8	70	39	44.2	68.5	65.3
SHNN	72.9	87.3	74.7	51.8	45	73.9	69.6
HG	76.5	85.6	72.5	49.1	40.25	72	68.2
HG+transformer	84	92	79	54	49	78.5	75.3

Table 7. Weighted F-score (%) of different edge-formation methods compared on BRIGHT dataset.

Model	DCIS	IC	PB	UDH	ADH	FEA	Weighted F-score
DSL	81.7	90.9	78.0	55.0	48.3	75.2	73.6
KNN on distance	81.1	90.4	77.8	56.8	57.7	75.9	74.9
KNN on node features	84.0	92	79	54	49	78.5	75.3

F-score by 4% whereas in the case of a heterogeneous convolutional network the addition of a transformer to heterogeneous convolutions (HG) gave an increase in the F-score of nearly 7%. It can be concluded that heterogeneous convolutions led to an increase in interactions between cell and tissue features such that the output of these convolutions has better features for the transformer to work on than that of normal sage convolutions. It is also to be noted that in our model we used just two heterogeneous convolutional layers as opposed to 6 graph sage convolutional layers in [12].

We tried three different graph formation methods for cell and tissue graph formation. The results are listed in Table 7. Our method of graph formation using kNN algorithm on node features was seen to perform better than all of the other three methods.

5   Conclusion

We introduced three novel architectures for histopathological image classification based on heterogeneous graph neural networks. Our research highlights the capability of heterogeneous graph convolutions in capturing the spatial as well as hierarchical relationship within the images, which allows it to surpass the performance of existing methods for histopathology image analysis. This emphasizes the significance of considering the relationship between cells and the surrounding tissue area for accurate cancer classification. Additionally, we established that the self-attention-based model outperforms the cross-attention-based model. We attribute this observation to the ability of self-attention to extract long-range dependencies within the graphs. Furthermore, we demonstrated the importance of analyzing the relationship between similar parts of the histopathology image, showcasing that constructing the graphs based on the similarity between node features yields superior results compared to other approaches, such as those based on spatial distance. However, we think that in the future both similarity and spatial distance should be combined for graph edge formation. It would also be good to explore novel techniques for enhancing graph convolutions in order to extract long-range dependencies within the graph more effectively. Additionally, there is a potential to develop innovative methods for graph pooling that minimize information loss. These directions of research would contribute to further advancements in histopathological image classification using heterogeneous graph neural networks.

References

1. Achanta, R., Shaji, A., Smith, K., Lucchi, A., Fua, P., Süsstrunk, S.: Slic superpixels compared to state-of-the-art superpixel methods. IEEE Trans. Pattern Anal. Mach. Intell. **34**(11), 2274–2282 (2012)
2. Anklin, V., et al.: Learning whole-slide segmentation from inexact and incomplete labels using tissue graphs. In: de Bruijne, M., et al. (eds.) MICCAI 2021. LNCS, vol. 12902, pp. 636–646. Springer, Cham (2021). https://doi.org/10.1007/978-3-030-87196-3_59
3. Aresta, G., et al.: BACH: grand challenge on breast cancer histology images. Med. Image Anal. **56**, 122–139 (2019)
4. Bai, J., Jiang, H., Li, S., Ma, X., et al.: NHL pathological image classification based on hierarchical local information and Googlenet-based representations. Biomed. Res. Int. **2019**, 1065652 (2019)
5. Benhammou, Y., Achchab, B., Herrera, F., Tabik, S.: BreakHis based breast cancer automatic diagnosis using deep learning: taxonomy, survey and insights. Neurocomputing **375**, 9–24 (2020)
6. Brancati, N., et al.: BRACS: a dataset for breast carcinoma subtyping in h&e histology images. In: Database 2022, baac093 (2022)
7. Chen, C.F.R., Fan, Q., Panda, R.: CrossViT: cross-attention multi-scale vision transformer for image classification. In: Proceedings of the IEEE/CVF International Conference on Computer Vision, pp. 357–366 (2021)
8. Chen, R.J., et al.: Pathomic fusion: an integrated framework for fusing histopathology and genomic features for cancer diagnosis and prognosis. IEEE Trans. Med. Imaging **41**(4), 757–770 (2020)
9. Graham, S., et al.: Hover-Net: simultaneous segmentation and classification of nuclei in multi-tissue histology images. Med. Image Anal. **58**, 101563 (2019)
10. Hamilton, W., Ying, Z., Leskovec, J.: Inductive representation learning on large graphs. In: Advances in Neural Information Processing Systems 30 (2017)
11. He, K., Zhang, X., Ren, S., Sun, J.: Deep residual learning for image recognition. In: Proceedings of the IEEE Conference on Computer Vision and Pattern Recognition, pp. 770–778 (2016)

12. Hou, W., Huang, H., Peng, Q., Yu, R., Yu, L., Wang, L.: Spatial-hierarchical graph neural network with dynamic structure learning for histological image classification. In: Wang, L., Dou, Q., Fletcher, P.T., Speidel, S., Li, S. (eds.) MICCAI 2022. LNCS, vol. 13432, pp. 181–191. Springer, Cham (2022). https://doi.org/10.1007/978-3-031-16434-7_18

13. Hou, W., Wang, L., Cai, S., Lin, Z., Yu, R., Qin, J.: Early neoplasia identification in Barrett's esophagus via attentive hierarchical aggregation and self-distillation. Med. Image Anal. **72**, 102092 (2021)

14. Jaume, G., Pati, P., Anklin, V., Foncubierta, A., Gabrani, M.: Histocartography: a toolkit for graph analytics in digital pathology. In: MICCAI Workshop on Computational Pathology, pp. 117–128. PMLR (2021)

15. Jia, Z., et al.: GraphSleepNet: adaptive spatial-temporal graph convolutional networks for sleep stage classification. In: IJCAI, vol. 2021, pp. 1324–1330 (2020)

16. Li, Y., Xie, X., Shen, L., Liu, S.: Reverse active learning based atrous densenet for pathological image classification. BMC Bioinformatics **20**(1), 1–15 (2019)

17. Lu, W., Graham, S., Bilal, M., Rajpoot, N., Minhas, F.: Capturing cellular topology in multi-gigapixel pathology images. In: Proceedings of the IEEE/CVF Conference on Computer Vision and Pattern Recognition Workshops, pp. 260–261 (2020)

18. Pati, P., et al.: HACT-Net: a hierarchical cell-to-tissue graph neural network for histopathological image classification. In: Sudre, C.H., et al. (eds.) UNSURE/GRAIL -2020. LNCS, vol. 12443, pp. 208–219. Springer, Cham (2020). https://doi.org/10.1007/978-3-030-60365-6_20

19. Pati, P., et al.: Hierarchical graph representations in digital pathology. Med. Image Anal. **75**, 102264 (2022)

20. Raju, A., Yao, J., Haq, M.M.H., Jonnagaddala, J., Huang, J.: Graph attention multi-instance learning for accurate colorectal cancer staging. In: Martel, A.L., et al. (eds.) MICCAI 2020. LNCS, vol. 12265, pp. 529–539. Springer, Cham (2020). https://doi.org/10.1007/978-3-030-59722-1_51

21. Vaswani, A., et al.: Attention is all you need. In: Advances in Neural Information Processing Systems 30 (2017)

22. Wu, Z., Pan, S., Chen, F., Long, G., Zhang, C., Philip, S.Y.: A comprehensive survey on graph neural networks. IEEE Trans. Neural Netw. Learn. Syst. **32**(1), 4–24 (2020)

23. Zhou, J., et al.: Graph neural networks: a review of methods and applications. AI Open **1**, 57–81 (2020)

24. Zhou, Y., Graham, S., Alemi Koohbanani, N., Shaban, M., Heng, P.A., Rajpoot, N.: CGC-NET: cell graph convolutional network for grading of colorectal cancer histology images. In: Proceedings of the IEEE/CVF International Conference on Computer Vision Workshops (2019)

25. Zielinska, H.A., et al.: Interaction between GRP78 and IGFBP-3 affects tumourigenesis and prognosis in breast cancer patients. Cancers **12**(12), 3821 (2020)

OCELOT 2023

SoftCTM: Cell Detection by Soft Instance Segmentation and Consideration of Cell-Tissue Interaction

Lydia Anette Schoenpflug$^{(\boxtimes)}$ [ID] and Viktor Hendrik Koelzer [ID]

Department of Pathology and Molecular Pathology, University Hospital and University of Zürich, Schmelzbergstrasse 12, 8091 Zürich, Switzerland
lydia.schoenpflug@usz.ch

Abstract. Detecting and classifying cells in histopathology H&E stained whole-slide images is a core task in computational pathology, as it provides valuable insight into the tumor microenvironment. In this work we investigate the impact of ground truth formats on the models performance. Additionally, cell-tissue interactions are considered by providing tissue segmentation predictions as input to the cell detection model. We find that a "soft", probability-map instance segmentation ground truth leads to best model performance. Combined with cell-tissue interaction and test-time augmentation our Soft Cell-Tissue-Model (Soft-CTM) achieves 0.7172 mean F1-Score on the Overlapped Cell On Tissue (OCELOT) test set, achieving the third best overall score in the OCELOT 2023 Challenge. The source code for our approach is made publicly available at https://github.com/lely475/ocelot23algo.

Keywords: histopathology image analysis · cell detection · deep learning · tumor microenvironment

1 Introduction

Cell detection and classification is a sub-task of Computational Pathology, which can be achieved through deep learning as shown in [2,5,8,10,14,15]. Jeongun Ryu and colleagues demonstrate in [11] that cell detection can benefit from considering cell-tissue relationships. They furthermore introduce the OCELOT dataset, which consists of 667 pairs of high resolution patches for cell detection in combination with lower resolution tissue patches, showing additional tissue context around the cell patch. The OCELOT 2023: Cell Detection from Cell-Tissue Interaction Challenge [9] motivates the development of cell detection algorithms that take the surrounding tissue context into account, based on the OCELOT dataset. In this paper, we present an approach to the OCELOT 2023 Challenge. First, we investigate the impact of different ground truth formats on the model performance. Second, we utilize the tissue segmentation annotation, by training a second model for tissue segmentation and providing its predictions as input to the cell detection model. Third, we utilize test-time augmentation

S.-A. Ahmadi and S. Pereira (Eds.): MICCAI 2023, LNCS 14373, pp. 109–122, 2024.
https://doi.org/10.1007/978-3-031-55088-1_10

(TTA), to further improve the model. Our final approach achieves the third best mean F1 Score of 0.7172 on the OCELOT test set.

2   Related Works

Deep learning approaches to cell detection in histopathology can be categorized into (1) Semantic segmentation-based approaches paired with instance extraction by either (a) postprocessing steps, such as a watershed transform [5,10], local maxima extraction [11] or morphological operations [15], or (b) an object detection network [2,8], and (2) pure object-detection approaches [14]. All approaches except [11] are trained on cell annotations only. In contrast, [11] motivates the consideration of tissue context for cell detection, demonstrating improved generalization for the OCELOT dataset. As the OCELOT dataset provides only cell centroid annotations, [11] translates them into a cell segmentation map by assigning pixels within a fixed radius of the nuclei centroid to the cells class. This potentially limits the model training, as only consideration of pixels around the nuclei centroid is rewarded. In contrast, [5,10] draw on full instance segmentation ground truths, enabling the consideration of all nuclei pixels. Furthermore, [10] translates the ground truth into a cell probability map instead of class labels to better reflect the cells blurry boundaries and enable a smoother prediction. This motivated us to extend [11], by enriching the ground truth formats from point annotations to instance segmentation maps and then investigating different ground truth formats for semantic segmentation.

3   Methods

In this section we describe the dataset and configuration for training a tumor segmentation model, as well as training a cell detection model on three different ground truth formats. Our proposed final workflow is a combined cell-tissue model (CTM) as described in Sect. 3.7.

3.1   Dataset

The OCELOT training dataset [11] consists of 400 pairs of cell and tissue patches of size 1024×1024, with a magnification of 50x and 12.5x respectively. Cell annotations are provided as nuclei centroid coordinates ("cell point annotation") for the classes tumor and background cells. Tissue annotations are provided as pixel-wise segmentation masks for the classes cancer area, background and unknown. We split the dataset into an internal training and validation set on a 80:20 split with stratified cell and tissue annotation and organ distribution (training set: 320 patch pairs from 138 WSIs, validation set: 80 patch pairs from 35 WSIs). The internal validation set is utilized for hyperparameter optimization. All further experiments are validated on the OCELOT validation set [11].

3.2   Model Architecture

Building on [11] we utilize a DeepLabv3+ segmentation model [4] with a ResNet50 [6] encoder pretrained on ImageNet[1] as shown in Fig. 1. The last ResNet50 convolutional block utilizes atrous convolutions with a rate of 2, to enable downsampling while preserving the input feature dimension. This is followed by an Atrous Spatial Pyramid Pooling (ASPP) block [3] and a decoder. The ASPP block extracts high-level-features at different downsampling rates to account for differing object scales. The decoder consists of two upsampling steps, connected by three convolutional layers which combine low-level features from the second ResNet50 layer with high-level features from the upsampled ASPP output. We utilize the same model architecture for cell detection and tissue segmentation.

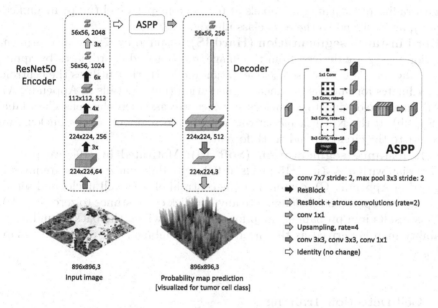

Fig. 1. Model architecture: DeepLabv3+ with ResNet50 Encoder, an Atrous Spatial Pyramid Pooling (ASPP) block and a Decoder with 2 upsampling and 3 convolutional blocks.

3.3   Tissue Segmentation Training

Training hyperparameters for the tissue segmentation model were based on [12]. The model was trained for 100 epochs on the internal training set to minimize a cross entropy loss with stochastic gradient descent (initial learning rate: 0.2,

---

[1] Pytorch default pretrained weights: https://pytorch.org/vision/stable/models/gene rated/torchvision.models.resnet50.html#torchvision.models.ResNet50_Weights, last accessed 24.11.2023.

Nesterov momentum: 0.9, weight decay: $5 \cdot 10^{-6}$, exponential learning rate decay: $\gamma = 0.97$) and a batch size of 8. The best model was selected based on the internal validation set. Training samples were oversampled to achieve a balanced amount of background and cancer pixels. The input images were augmented by re-scaling in the range $\pm 10\%$, random crop to 896×896 pixels, flip, rotation by $90°$, $180°$ or $270°$ and a channel-wise brightness and contrast variation by $\pm 20\%$. Each augmentation is applied with a probability of 70%.

3.4   Cell Detection Ground Truth Formats

The sparse cell point annotations require translation into segmentation annotations. We investigate the following ground truth formats:

1. **Circle:** Identical to [11], all pixels in a circle, centered on the cell coordinates with radius of $1.4\,\mu m$ ($r = 7$ pixels at 0.2 microns-per-pixel (mpp) magnification), are assigned to the cells class id (Fig. 2a).
2. **Hard instance segmentation (Hard IS):** Inspired by [7] instance segmentation masks are derived from the image and centroid coordinates by applying NuClick[2], a CNN-based segmentation model [1] that utilizes the centroid coordinates for nucleus instance segmentation (more details in Appendix A). All pixels belonging to a nucleus instance are assigned the cells class label (Fig. 2b). If the NuClick model was not able to segment a cells nucleus, we revert to the circle ground truth format.
3. **Soft instance segmentation (Soft IS):** Motivated by [10], we place a Gaussian with $\sigma = 3\,\mu m$ (15 pixels at 0.2 mpp, different σ values are investigated in Appendix C), centered at the centroid of each cell nuclei and set all background pixels not belonging to any NuClick cell instance to zero (Fig. 2c). This results in a probability map for each class, where the background probability map is derived as the inverse of the combined cell probability maps $y_{bg} = 1 - \sum_c y_c$.

3.5   Cell Detection Training

Model hyperparameters such as learning rate, optimizer, architecture and loss function were selected based on performance on the internal validation set. For the learning rate we considered values in the range $[5 \cdot 10^{-5}, 2 \cdot 10^{-3}]$. The model was trained on the OCELOT training set in a k-fold manner with fixed hyperparameters ($k = 5$), resulting in an 80:20 training to validation split for each iteration. The five trained models were combined by using the averaged sum of their predictions. The cell detection models were trained for 150 epochs with a weighted Adam optimizer (learning rate: $8 \cdot 10^{-4}$) and a batch size of 32.

---

[2] Publicly available at https://github.com/navidstuv/NuClick, last accessed 24.11.2023.

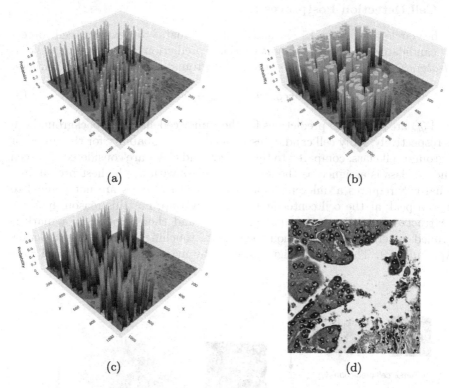

Fig. 2. Example of different ground truth formats, visualization of the tumor cell class probability map: (a) Circle, (b) Hard IS, (c) Soft IS, (d) Original image with cell point annotation (blue: tumor cells, yellow: background cells) (Color figure online)

The learning objective is minimizing a Dice loss [13] for the circle and hard IS ground truth format:

$$Generalized\ Dice\ Loss = 1 - 2 \cdot \frac{\sum_c w_c \cdot \sum_{i=1}^{N} y_{i,c} \cdot \hat{y}_{i,c}}{\sum_c w_c \cdot \sum_{i=1}^{N} y_{i,c} + \hat{y}_{i,c}}, \quad w_c = \frac{1}{\sum_i y_{i,c}} \quad (1)$$

where $y_{i,c}$ is the segmentation ground truth and $\hat{y}_{i,c}$ is the segmentation prediction for pixel i and class c, with a weighting of w_c for each class. The soft IS format poses a pixel-wise regression problem, for this reason we utilize a weighted mean square error (MSE) loss:

$$Weighted\ MSE\ Loss = \sum_c w_c \cdot \frac{1}{N} \sum_{i=1}^{N} (y_{i,c} - \hat{y}_{i,c})^2, \quad w_c = \frac{\sum_i y_i}{\sum_i y_{i,c}} \quad (2)$$

Training samples were over-sampled based on the presence of background and tumor cells in each sample, to achieve a balanced number of tumor and background cell instances. The same augmentation methods as for the tumor segmentation model were applied.

3.6   Cell Detection Postprocessing

For the circle and soft IS ground truth formats we extract cell detection candidates from the segmentation prediction by applying *skimage.feature.peak_local_max* on the blurred foreground prediction (Fig. 3), where:

$$foreground = \hat{y}_{tc} + \hat{y}_{bc} \tag{3}$$

$\hat{y}_{tc}$ and $\hat{y}_{bc}$ are the model predictions for the tumor cell class and background cell class respectively. Only cell candidates with a larger probability for the tumor or background cell class, compared to the background class, are considered. The cell candidate class is assigned as the foreground class with the highest probability. The hard IS requires a different approach, as cell instances are not trained to express a peak at the cell center and tend to overlap. For this reason, markers are extracted from the foreground prediction and then applied in a marker-controlled watershed segmentation to separate touching instance (more details in Appendix B). The cell is assigned a class by majority vote of its pixel class predictions.

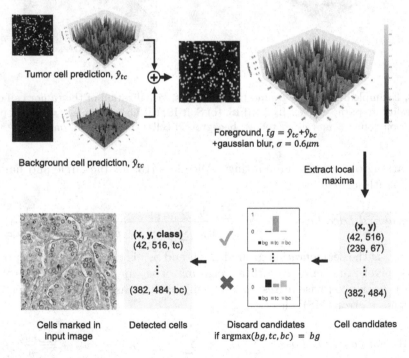

Fig. 3. Postprocessing for circle and soft IS ground truth format, tc: tumor cell class, bc: background cell class

3.7   Combined Cell-Tissue Model

We combine the cell detection and tissue segmentation models, by providing and re-training the cell detection model with both the cell patch and the cropped and upsampled tissue segmentation prediction as input (Fig. 4). The cell detection model ground truth configuration is chosen based on the best performance on the OCELOT validation set. Additionally, we evaluated the effect of geometrical TTA for more robust predictions, consisting of all 8 possible rotation and flip combinations. TTA was applied for both models.

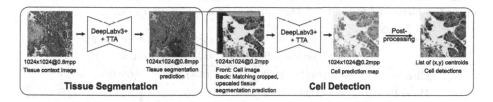

Fig. 4. Combined cell-tissue model for cell detection: First, tissue segmentation is performed on a image showing surrounding tissue context of the cell patch. The tissue segmentation prediction is cropped and upsampled to match the cell patch, together forming the input to the cell detection model. Second, cell detection is performed resulting in a cell prediction map, which is postprocessed to extract class-wise cell centroid coordinates.

4   Results

We evaluate the performance of the cell detection models trained on the three ground truth formats, as well as the performance of the CTM. Lastly, we investigate the effect of adding TTA.

4.1   Main Findings

Table 1 details the mean F1 score for the three ground truth formats. The best performing ground truth format is the soft IS across all sets. It is notable that the OCELOT test set shows a performance decrease from circle to hard IS ground truth. This is not the case for the internal validation and OCELOT validation set, but nevertheless highlights that while the hard IS ground truth increases the number of cell class pixels, this does not necessarily translate into a better model performance. In contrast, utilizing the soft IS leads to a performance increase on all sets, possibly due to rewarding a local maxima, clearer cell boundaries and a simplified postprocessing. As the soft IS showed highest performance among the three ground truth format models, we train the CTM with the soft IS ground truth for cell detection, further referred to as the SoftCTM. The Soft-CTM shows an increased mean F1 for the internal validation and OCELOT test set, but reduced mean F1 for the OCELOT validation set. However, combining the SoftCTM with test-time augmentation leads to the overall best score on the validation and test set. The need for TTA to improve performance on the validation set, when utilizing the SoftCTM, indicates that the models are to a certain extent sensitive to geometric variations. Yet, this is not observed on the internal validation and test set (Table 2).

Table 1. Comparison of mean F1 scores for cell detection. The CTM was trained with the soft IS ground truth, as this showed the highest performance among the ground truth formats and is therefore denoted as SoftCTM. The highest score for the three ground truth formats is underlined, the overall highest score is marked in **bold**. We report the mean and std of the 5-Fold internal cross validation runs.

	5-Fold Internal validation	OCELOT validation	OCELOT test
Circle	.5647±.0262	.6781	.6599
Hard IS	.6029±.0442	.6826	.6516
Soft IS	.6494±.0302	.6937	.6777
SoftCTM	.6842±.0238	.6875	.7090
SoftCTM + TTA	**.6950±.0245**	**.7046**	**.7172**

Table 2. Comparison of mean F1 scores for tissue segmentation. We report the performance on the fixed internal validation set, as well as the OCELOT validation and test set. The highest score for each set is marked in **bold**.

	Internal validation	OCELOT validation	OCELOT test
Tissue segmentation	.9131	.8511	.8927
Tissue segmentation + TTA	**.9174**	**.8571**	**.8951**

4.2 Organ-Wise Results

Table 3 provides insight into the per-organ mean F1 for cell detection and tissue segmentation on the OCELOT test set, for the validation set organ-wise performance is reported in Appendix D. For kidney, endometrium, stomach and head-neck we observe the same tendencies as in the full set. In contrast, for prostate the three ground truth formats show only very little difference in performance, with the circle ground truth outperforming the others, while using the SoftCTM and TTA improves model performance. This is not the case for bladder, where the SoftCTM results in lower performance. We suspect this might be related to the less accurate tissue prediction for bladder, with approximately 0.84 F1-Score in contrast to > 0.9 for the majority of organs (Table 4). However, prostate samples show an even lower tissue segmentation performance, yet utilizing the SoftCTM has a positive effect on the cell detection performance.

We investigated this further by using the tissue segmentation ground truth as input to the SoftCTM[3], we refer to this mode as the Tissue-label leaking model (TLLM). The largest improvement of utilizing the TLLM is observed for bladder samples with +10% mean F1 score, which confirms that faulty tissue segmentation predictions lead to degraded cell detection performance. At the same time, we note that using the TLLM lead to a slight performance decrease compared to the SoftCTM for organs endometrium, stomach and head-neck, indicating that the tissue segmentation model appears to better capture the tissue composition than the ground truth for this subset.

Table 3. Comparison of per-organ mean F1 scores on the OCELOT test set for cell detection. The CTM trained with the soft IS ground truth is denoted as SoftCTM. The tissue-label leaking model is denoted as TLLM. The highest score for the three ground truth formats is <u>underlined</u>, the overall highest score is marked in **bold**.

	all n = 130	kidney n = 41	endometrium n = 25	bladder n = 26	prostate n = 16	stomach n = 12	head-neck n = 10
Circle	.6599	.6457	.6791	.6317	<u>.6276</u>	.6573	.6547
Hard IS	.6516	.6551	.6486	.6276	.6228	.6593	.6624
Soft IS	<u>.6777</u>	<u>.6608</u>	<u>.7087</u>	**.6515**	.6220	<u>.6911</u>	<u>.6966</u>
SoftCTM	.7090	.7208	.7527	.6240	.6416	.7126	.7457
SoftCTM + TTA	**.7172**	**.7323**	**.7560**	.6291	**.6525**	**.7360**	**.7474**
TLLM	.7269	.7409	.7225	.7259	.6921	.7092	.7314
TLLM + TTA	.7315	.7469	.7210	.7300	.7063	.7219	.7338

Table 4. Comparison of per-organ mean F1 scores on the OCELOT test set for tissue segmentation. The tissue segmentation model is denoted as TSM. The highest score for each organ is marked in **bold**.

	all n = 130	kidney n = 41	endometrium n = 25	bladder n = 26	prostate n = 16	stomach n = 12	head-neck n = 10
TSM	.8927	.9241	.9062	.8358	.8119	.9014	**.8863**
TSM + TTA	**.8951**	**.9282**	**.9088**	**.8383**	**.8178**	**.9035**	.8774

5   Conclusion

Among the studied ground truth formats, extending the cell ground truth with the NuClick segmentation prediction and utilizing soft instance segmentation maps lead to the largest improvement in model generalization. By further combining it with tissue predictions and test-time augmentation we achieve 0.7172 mean F1 Score on the OCELOT test set. This supports the assumption that

---

[3] The tissue segmentation prediction was kept for all ground truth pixels of class Unknown and only replaced for the Background and Cancer Area pixels.

cell detection can benefit from considering the tissue context. As the OCELOT validation and test set originate from similar domains and are limited in size, future work could be conducted on evaluating the models usability on a larger cohort with respect to downstream tasks such as cell content estimation.

Appendix

We provide the following supplementary material:

– Description of segmentation ground truth generation with NuClick
– Postprocessing steps when using the hard segmentation ground truth format
– An Ablation study considering different σ values for the soft IS ground truth
– Organ-wise performance on OCELOT validation set

A   Segmentation Ground Truth Generation with NuClick

We utilize NuClick [1], a pretrained nucleus, cell and gland segmentation model[4], to extend the cell annotations from centroid coordinates to segmentation maps, as visualized in Fig. 5. It relies on a guiding signal, in our case the cell point annotations, together with the input image for instance segmentation. Pretrained weights are only available for nuclei segmentation[5], for this reason we extend the ground truth to a nuclei segmentation mask instead of a cell segmentation mask.

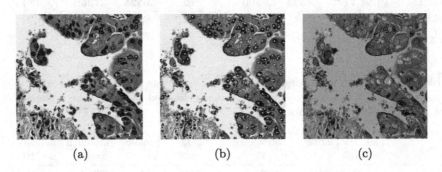

(a)   (b)   (c)

Fig. 5. (a) Original image, (b) Visualized point annotations, (c) Nuclick prediction

The NuClick scripts required slight adaptation for our use case:

– Read in cell points annotations in csv file format instead of mat.
– Read in images in jpg format instead of tif.
– Update tensorflow functions to avoid usage of deprecated functions.

The updated scripts are made public as a Github repository[6].

[4] Publicly available at https://github.com/navidstuv/NuClick, last accessed 24.11.2023.
[5] https://drive.google.com/file/d/1MGjZs_-2Xo1W9NZqbq_5XLP-VbIo-ltA/view, last accessed: 24.11.2023.
[6] NuClick repository with adaptations: https://github.com/lely475/NuClick.

B   Postprocessing for the Hard IS Ground Truth Format

For the hard IS ground truth format, cell detection candidates are extracted by first combining the tumor and background cell prediction into a foreground prediction. The Otsu threshold is then applied on the foreground, resulting in a binary foreground map, from which holes and small objects are removed using the *skimage.morphology.remove_small_objects* and *scipy.ndimage.binary_fill_holes* functions. Next, we calculate the Euclidean Distance Transform (EDT) on the binary image, revealing the distance of each pixel to the nearest background pixel and, consequently, highlighting potential cell instances as peaks. To identify these peaks, we employ *skimage.feature.peak_local_max* function, assigning each peak a unique identifier. These identified points serve as markers in a marker-controlled watershed technique, facilitating the separation of connected objects within the binary foreground map. The inverse EDT (-EDT) is used as a cell border indicator. The final cell candidates are determined by locating the center of mass within each segmented instance of the watershed segmentation. The cell class is assigned as the class with the majority of class pixels in each cell instance (Fig. 6).

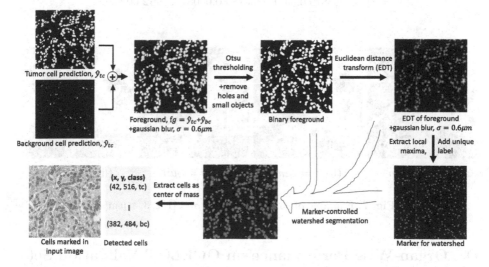

Fig. 6. Postprocessing for hard IS ground truth format, tc: tumor cell class, bc: background cell class

C   Ablation Study: Different σ Values for Soft IS Ground Truth

The $\sigma = 3\,\mu\mathrm{m}$ value for soft IS was originally chosen based on a visual review. To investigate, whether there would be a more suitable σ value, we conducted an ablation study with $\sigma = [1, 2, 3, 4]\,\mu\mathrm{m}$, analyzing the performance of the soft IS model. Figure 7 shows an example of the soft IS mask for all considered σ values. The internal train and validation set were used for training and evaluation. The highest mean F1-Score is achieved for $\sigma = 2\,\mu\text{m}$, while our initial choice $\sigma = 3\,\mu\text{m}$ shows a slightly lower performance by -0.6%. Notably, the choice of σ appears to be a compromise between precision and recall (Table 5). While lower σ lead to a more precise cell detection, this comes at the cost of a higher number of missed cells (false negatives). The opposite is the case for larger σ values. Overall, while the difference in F1-Score for $\sigma = 2\,\mu\text{m}$ and $\sigma = 3\,\mu\text{m}$ is minor, utilizing $\sigma = 2\,\mu\text{m}$ for future work is recommended.

Table 5. Performance comparison of different $\sigma = [1, 2, 3, 4]\,\mu\text{m}$ for the soft IS probability map on the internal validation set. We report the mean and std of 3 runs.

	F1-Score	Precision	Recall
$\sigma = 1\,\mu\text{m}$	.5939±0.0044	**.7695**±0.0092	.4760±0.0035
$\sigma = 2\,\mu\text{m}$	**.6709**±0.0013	.7234±0.0088	.6266±0.0083
$\sigma = 3\,\mu\text{m}$	.6649±0.0026	.6894±0.0048	.6424±0.0053
$\sigma = 4\,\mu\text{m}$	.6587±0.0031	.6618±0.0008	**.6562**±0.0056

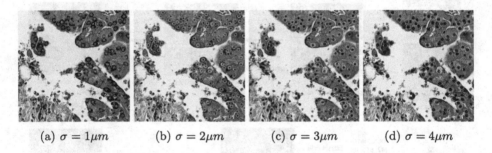

(a) $\sigma = 1\mu m$ (b) $\sigma = 2\mu m$ (c) $\sigma = 3\mu m$ (d) $\sigma = 4\mu m$

Fig. 7. Example of soft IS mask for $\sigma = [1, 2, 3, 4]\,\mu\text{m}$

D   Organ-Wise Performance on OCELOT Validation Set

Tables 6 and 7 detail the results for each approach on the OCELOT validation set. Notably, while there is a trend for the soft IS, SoftCTM and TTA to improve performance for the majority of organs, there are also multiple organs for which one or multiple of these trends are not present. This highlights the challenge of training a model which can generalize well to different organs.

Table 6. Comparison of per-organ mean F1 scores on the OCELOT validation set for cell detection. The CTM trained with the soft IS ground truth is denoted as SoftCTM. The highest score for the three ground truth formats is underlined, the overall highest score is marked in **bold**.

	all n = 137	kidney n = 41	endometrium n = 29	bladder n = 29	prostate n = 17	stomach n = 12	head-neck n = 9
Circle	.6781	.6580	.7167	.5824	.6174	.6148	.5470
Hard IS	.6826	.7101	.7094	.5852	.6076	.6984	.5401
Soft IS	.6937	.6934	.7437	.5818	.6136	.7041	.5682
SoftCTM	.6875	.6432	.7258	.6363	.6328	.7333	.5832
SoftCTM + TTA	**.7046**	.6861	.7397	**.6364**	**.6385**	**.7412**	**.5848**

Table 7. Comparison of per-organ mean F1 scores on the OCELOT validation set for tissue segmentation. The tissue segmentation model is denoted as TSM. The highest score for each organ is marked in **bold**.

	all n = 130	kidney n = 41	endometrium n = 25	bladder n = 26	prostate n = 16	stomach n = 12	head-neck n = 10
TSM	.8511	.8103	.9258	**.8270**	.7914	.6187	**.7752**
TSM + TTA	**.8571**	**.8281**	**.9307**	.8186	**.7970**	**.6243**	.7693

References

1. Alemi Koohbanani, N., Jahanifar, M., Zamani Tajadin, N., Rajpoot, N.: Nuclick: a deep learning framework for interactive segmentation of microscopic images. Med. Image Anal. **65**, 101771 (2020). https://doi.org/10.1016/j.media.2020.101771, https://www.sciencedirect.com/science/article/pii/S1361841520301353

2. Bancher, B., Mahbod, A., Ellinger, I., Ecker, R., Dorffner, G.: Improving mask R-CNN for nuclei instance segmentation in hematoxylin & eosin-stained histological images. In: Atzori, M., et al. (eds.) Proceedings of the MICCAI Workshop on Computational Pathology. Proceedings of Machine Learning Research, vol. 156, pp. 20–35. PMLR (2021). https://proceedings.mlr.press/v156/bancher21a.html

3. Chen, L.C., Papandreou, G., Kokkinos, I., Murphy, K., Yuille, A.L.: DeepLab: Semantic image segmentation with deep convolutional nets, Atrous convolution, and fully connected CRFs. IEEE Trans. Pattern Anal. Mach. Intell. **40**(4), 834–848 (2018). https://doi.org/10.1109/TPAMI.2017.2699184

4. Chen, L.C., Zhu, Y., Papandreou, G., Schroff, F., Adam, H.: Encoder-decoder with Atrous separable convolution for semantic image segmentation. In: Ferrari, V., Hebert, M., Sminchisescu, C., Weiss, Y. (eds.) Computer Vision - ECCV 2018, pp. 833–851. Springer International Publishing, Cham (2018)

5. Graham, S., et al.: Hover-net: simultaneous segmentation and classification of nuclei in multi-tissue histology images. Med. Image Anal. **58**, 101563 (2019). https://doi.org/10.1016/j.media.2019.101563, https://www.sciencedirect.com/science/article/pii/S1361841519301045

6. He, K., Zhang, X., Ren, S., Sun, J.: Deep residual learning for image recognition. In: Proceedings of the IEEE Conference on Computer Vision and Pattern Recognition, pp. 770–778 (2016)

7. Jahanifar, M., et al.: Stain-robust mitotic figure detection for the mitosis domain generalization challenge. In: Aubreville, M., Zimmerer, D., Heinrich, M. (eds.) MICCAI 2021. LNCS, vol. 13166, pp. 48–52. Springer, Cham (2022). https://doi.org/10.1007/978-3-030-97281-3_6
8. Liu, D., et al.: Nuclei segmentation via a deep panoptic model with semantic feature fusion. In: IJCAI, pp. 861–868 (2019)
9. Lunit Inc.: OCELOT 2023: Cell Detection from Cell-Tissue Interaction - Grand Challenge – ocelot2023.grand-challenge.org (2023). https://ocelot2023.grand-challenge.org/. Accessed 29 Sep 2023
10. Naylor, P., Laé, M., Reyal, F., Walter, T.: Segmentation of nuclei in histopathology images by deep regression of the distance map. IEEE Trans. Med. Imaging **38**(2), 448–459 (2019). https://doi.org/10.1109/TMI.2018.2865709
11. Ryu, J., et al.: Ocelot: overlapped cell on tissue dataset for histopathology. In: Proceedings of the IEEE/CVF Conference on Computer Vision and Pattern Recognition, pp. 23902–23912 (2023)
12. Schoenpflug, L.A., Lafarge, M.W., Frei, A.L., Koelzer, V.H.: Multi-task learning for tissue segmentation and tumor detection in colorectal cancer histology slides. arXiv preprint arXiv:2304.03101 (2023)
13. Sudre, C.H., Li, W., Vercauteren, T., Ourselin, S., Jorge Cardoso, M.: Generalised dice overlap as a deep learning loss function for highly unbalanced segmentations. In: Cardoso, M.J., et al. (eds.) DLMIA/ML-CDS -2017. LNCS, vol. 10553, pp. 240–248. Springer, Cham (2017). https://doi.org/10.1007/978-3-319-67558-9_28
14. Sun, Y., Huang, X., Zhou, H., Zhang, Q.: SRPN: similarity-based region proposal networks for nuclei and cells detection in histology images. Med. Image Anal. **72**, 102142 (2021). https://doi.org/10.1016/j.media.2021.102142, https://www.sciencedirect.com/science/article/pii/S1361841521001882
15. Zeng, Z., Xie, W., Zhang, Y., Lu, Y.: RIC-Unet: an improved neural network based on Unet for nuclei segmentation in histology images. IEEE Access **7**, 21420–21428 (2019). https://doi.org/10.1109/ACCESS.2019.2896920

Detecting Cells in Histopathology Images with a ResNet Ensemble Model

Maxime W. Lafarge$^{(\boxtimes)}$ and Viktor Hendrik Koelzer

Department of Pathology and Molecular Pathology, University Hospital Zurich,
University of Zurich, Zurich, Switzerland
`maxime.lafarge@usz.ch`

Abstract. This paper presents a candidate solution that was submitted to the test phase of the OCELOT2023 challenge. This solution is based on an ensemble of trained fully convolutional ResNet-50 models that were developed using only the small field-of-view (FoV) images with cell-level annotations of the challenge dataset (the large FoV images with tissue-level annotations were not used). The submitted model achieved a F_1-score of 0.673 on the evaluation set of the validation phase. The code to run our submitted trained model is available at: https://github.com/CTPLab/OCELOT2023_P4ResNet50.

Keywords: cell detection · cell classification · computational pathology · deep learning · ensemble modeling · OCELOT2023

1 Introduction

Automated detection and classification of single cells in hematoxylin-and-eosin histopathology images is a useful task with high-potential applications in downstream analysis [6]. To further support the development of efficient machine learning solutions to solve this task, the organizers of the OCELOT2023 challenge released the OCELOT dataset [13].

The OCELOT dataset consists of annotated images of tissue regions (extracted from TCGA [5]) from six different organs at two magnification levels (small field-of-view (FoV) images with cell-level annotations at magnification 40× that overlap large FoV images with tissue-level annotations at magnification 10×). As the authors of [13] previously showed that better performance for this detection/classification task can be achieved when using both magnification levels for training and inference, the challenge encourages participants to develop solutions in the same conditions.

The organizers trained a baseline model [1] based on a U-Net architecture [12] using only the small FoV cell-level annotated part of the challenge dataset. We argue that this control experiment is highly valuable to assess the gain of performance of solutions that use the full challenge dataset, and this is why we decided to conduct a similar *control experiment* by investigating a different algorithm that also relies exclusively on the small FoV cell-level annotated part of the challenge dataset.

S.-A. Ahmadi and S. Pereira (Eds.): MICCAI 2023, LNCS 14373, pp. 123–129, 2024.
https://doi.org/10.1007/978-3-031-55088-1_11

For this purpose, we trained and validated three models based on a customized ResNet-50 convolutional neural network, and combined them into an ensemble model that constitutes our final submitted solution. Each model was trained and validated using a different fold of the OCELOT training dataset. This work is motivated by the need for straight-forward pipelines that rely on minimal input, which can be important in contexts with limited computational ressources to enable the analysis of large datasets of whole slide images at a reduced computational cost. We thus used a data-centric approach to address this problem by fixing the model architecture at the beginning of the development phase and by focusing on how to best use the available data.

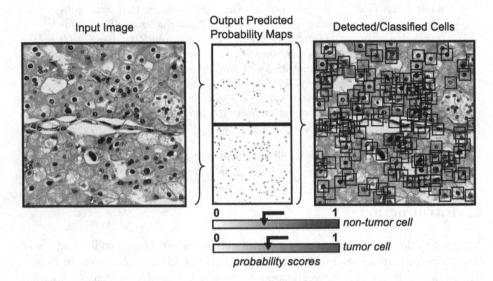

Fig. 1. Example of the application of a trained customized ResNet-50 to an unseen image from the OCELOT validation dataset to produce class-wise probability maps (middle-top: non-tumor cells, middle-bottom: tumor cells). Candidate cells are identified as local maxima in these probability maps. Class-and-organ-wise cutoff values selected to optimize performance in the internal validation sets (represented by arrows) are used to detect and classify cells as shown on the right-most image.

2 Model Architecture

Our approach is based on training a customized ResNet architecture [9] with 50 layers. We made the model invariant to 90-degree rotation via the use of group-equivariant convolution layers as described in [10]. The architecture was designed to process input image patches of size 130×130px and to output a softmax-activated 3-class probability vector that predicts the class of the center point of the input (either background, non-tumor (normal) cell or tumor cell). We made the model fully convolutional by using valid-padding convolutions and by using cropping in skip connections of residual blocks as motivated by the approach described in [11]. The implementation of the resulting inference pipeline can be found at: https://github.com/CTPLab/OCELOT2023_P4ResNet50.

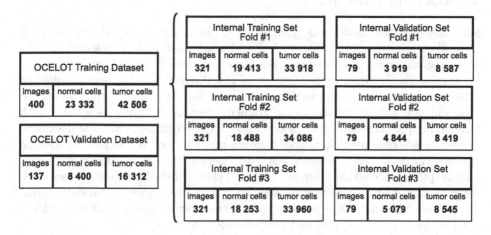

Fig. 2. Summary of the dataset used for the development of the submitted ensemble model. Three training-validation splits were derived from the OCELOT training dataset and were used for training and model selection. The validation sets do not overlap across folds. The OCELOT validation dataset was not available during the development phase and was used to further validate models after the challenge.

3   Dataset Preparation

Given the small FoV cell-level annotated images, the true class of pixels were defined either as the class of the strictly closest ground-truth annotated cell center (either non-tumor (normal) cell or tumor cell) within a radius of 10px, or as background otherwise. In the case a pixel was equidistant to two or more ground-truth center points, it was labeled as background. We created three folds of internal training-validation splits (distribution 80–20) by partitioning the OCELOT training dataset at image-level while ensuring a similar distribution of organs and classes across the splits. An overview of this partitioning is described in Fig. 2. Based on the fine-grained hard negative mining procedure described in [11], we trained a first instance of the model using the whole dataset, and then applied the resulting trained model $\mathcal{M}_0$ to the whole dataset to identify background regions that were correctly classified with high confidence. These background regions were then excluded from the training sets such that there was a 2:1 ratio between the overall number of background patches and number of patches centered on normal/tumor cells. As an example, to train our final model with the internal training set of Fold#1 (Fig. 2), we used patches centered on

19 413 normal cells, 33 918 tumor cells, and the 106 662 background patches in this training set that are not centered on annotated normal/tumor cells and that obtained the lowest probability scores based on $\mathcal{M}_0$.

4   Training Procedure

For each fold, the model was trained using image patches randomly sampled from the training split, while excluding background regions selected by the hard negative mining procedure. We used batches of size 64 while ensuring an approximately equal representation of the three classes via oversampling of minority classes, and optimized the model via minimization of the categorical cross-entropy loss, using stochastic gradient descent (learning rate 0.01) with momentum (coefficient 0.9). We used weight decay (coefficient 10^{-4}) and a cosine annealing schedule over the whole training procedure of 10^6 iterations. During training, we periodically assessed the model performance using the internal validation split and saved the state of the model that achieved the best performance for inference. Data augmentation was used with the following random transformations: horizontal flip (probability 0.5), random rotation (multiple of 15°), random zoom (range ±20%) and random color jitter.

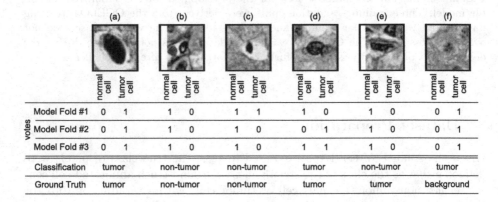

	(a) normal cell	(a) tumor cell	(b) normal cell	(b) tumor cell	(c) normal cell	(c) tumor cell	(d) normal cell	(d) tumor cell	(e) normal cell	(e) tumor cell	(f) normal cell	(f) tumor cell
Model Fold #1	0	1	1	0	1	1	1	0	1	0	0	1
Model Fold #2	0	1	1	0	1	0	0	1	1	0	0	1
Model Fold #3	0	1	1	0	1	0	1	1	1	0	0	1
Classification	tumor		non-tumor		non-tumor		tumor		non-tumor		tumor	
Ground Truth	tumor		non-tumor		non-tumor		tumor		tumor		background	

Fig. 3. Example of the ensemble voting procedure used to classify candidate detected objects based on the individual predictions of three trained models developed on separate data folds. The resulting ensemble model enables the identification of cases with robust classification (a,b,e,f) and uncertain classification (c,d).

5   Inference Pipeline

For inference, the fully convolutional structure of the model enabled its direct application to images of size 1024×1024px and was used produce two probability maps for non-tumor cells and tumor cells. For each model, we averaged the predicted probability maps of flipped input images as test time augmentation. Then, all local maxima (within a radius of 16px) in a given probability map were extracted, and a local maximum location was considered as a positive detection if its probability was above a cutoff value that had been selected to maximize the F_1-score on the validation split. An example of the application of a trained model to an unseen image from the challenge validation dataset is shown in Fig. 1. To avoid any biased selection of cutoff values, different optimal cutoff values were selected for each class-organ pair.

Then, all the points detected by the three trained models were combined, and adjacent points (distance lower than 16px) were merged: each resulting merged point corresponds to a candidate detected object. During merging, the resulting merged points were given a vote for one of the two classes by each of their source points. Finally, points were classified according to the class that obtained a strict majority of votes [7]. To ensure robust predictions, we classified candidate detected objects that obtained a single vote as background, and ambiguous objects with equal number of votes for both classes (greater than or equal to two) as tumor cells. Examples of classified objects resulting from this ensemble voting procedure are shown in Fig. 3.

Table 1. Comparison of F_1-scores for detection/classification of non-tumor/background cells (BC-F1) and tumor cells (TC-F1) of different models.

	Internal Validation		
	BC-F1	TC-F1	mean-F1
Fold #1	0.617	0.732	0.675
Fold #2	0.634	0.703	0.669
Fold #3	0.657	0.675	0.667
	Challenge Validation Phase		
	BC-F1	TC-F1	mean-F1
Ensemble Model (submitted)	0.611	0.735	**0.673**
U-Net baseline [1,2]	0.589	0.723	0.656
	Post-Challenge OCELOT Validation		
	BC-F1	TC-F1	mean-F1
Fold #1	0.580	0.721	0.651
Fold #2	0.599	0.749	0.674
Fold #3	0.598	0.721	0.660

6   Results and Discussion

Performance of the developed models on our internal validation splits and on the OCELOT validation set are summarized in Table 1. As a result, we observed that our solution based on an ensemble of three trained models based on the same ResNet-50 architecture can achieve a slightly higher mean F1-score (+0.017) in comparison with the baseline algorithm of the challenge [1]. This suggests that there is a margin to improve the classification performance of standard models that are trained using only cell-level annotations.

These results also suggest that up-sampling layers (as required in the approaches based on a U-Net architecture, but absent in the proposed solution) might not be essential to achieve similar performances in the context of the task of this challenge.

In the comparative analysis, the ensemble model had a better mean-F_1-score on the OCELOT validation set than two of the three individual models, which confirms that ensembling is a sensible approach to ensure optimal performance as opposed to arbitrarily choosing a single model. In these experiments, selecting a model based on the highest mean-F_1-score on the internal validation set would have led to sub-optimal mean-F_1-score on the OCELOT validation set. This result is in line with the previously reported efficiency of ensembling in past cell-level classification challenges [3,4,8].

The leaderboard of the validation phase of the OCELOT challenge [1] reveals that higher classification performance can be achieved for this task in comparison with our solution. As concluded in [13], this gap of performance can be explained by the fact that our approach did not use the large FoV images with tissue-level annotations, which were shown to be necessary to achieve optimal classification performance for this task. In future work we will aim at investigating and explaining what information is used by models trained with large FoVs to correctly classify cells that are misclassified by the proposed model.

References

1. OCELOT 2023: Cell detection from cell-tissue interaction (2023). https://ocelot2023.grand-challenge.org/
2. OCELOT 2023: U-Net example (2023). https://github.com/lunit-io/ocelot23algo/tree/main/user/unet_example
3. Aubreville, M., et al.: Mitosis domain generalization in histopathology images - the MIDOG challenge. Med. Image Anal. **84**, 102699 (2023)
4. Aubreville, M., et al.: Domain generalization across tumor types, laboratories, and species-insights from the 2022 edition of the mitosis domain generalization challenge (2023). arXiv preprint arXiv:2309.15589
5. Network, C.G.A.: Comprehensive molecular portraits of human breast Tumours. Nature **490**(7418), 61–70 (2012)
6. Diao, J.A., et al.: Human-interpretable image features derived from densely mapped cancer pathology slides predict diverse molecular phenotypes. Nat. Commun. **12**, 1613 (2021)
7. Ganaie, M.A., Hu, M., Malik, A., Tanveer, M., Suganthan, P.: Ensemble deep learning: a review. Eng. Appl. Artif. Intell. **115**, 105151 (2022)
8. Graham, S., et al.: CoNIC challenge: Pushing the frontiers of nuclear detection, segmentation, classification and counting (2023). arXiv preprint arXiv:2303.06274

9. He, K., Zhang, X., Ren, S., Sun, J.: Identity mappings in deep residual networks. In: European Conference on Computer Vision (ECCV), pp. 630–645 (2016)
10. Lafarge, M.W., Bekkers, E.J., Pluim, J.P., Duits, R., Veta, M.: Roto-translation equivariant convolutional networks: application to histopathology image analysis. Med. Image Anal. **68**, 101849 (2021)
11. Lafarge, M.W., Koelzer, V.H.: Fine-grained hard-negative mining: generalizing mitosis detection with a fifth of the MIDOG 2022 dataset. In: MICCAI Challenge on Mitosis Domain Generalization (2022)
12. Ronneberger, O., Fischer, P., Brox, T.: U-Net: convolutional networks for biomedical image segmentation. In: Proceedings of the International Conference on Medical Image Computing and Computer-Assisted Intervention (MICCAI), pp. 234–241 (2015)
13. Ryu, J., et al.: OCELOT: overlapped cell on tissue dataset for histopathology. In: Proceedings of the IEEE/CVF Conference on Computer Vision and Pattern Recognition (CVPR) (2023)

Enhancing Cell Detection via FC-HarDNet and Tissue Segmentation: OCELOT 2023 Challenge Approach

Yu-Wen Lo$^{(\boxtimes)}$ and Ching-Hui Yang

Department of Computer Science, National Tsing Hua University, Hsinchu, Taiwan
{wagw1014,hui09080729}@gapp.nthu.edu.tw
https://github.com/YuWenLo/OCELOT2023

Abstract. Accurate cell detection is essential for numerous applications in biomedical research, including cancer diagnosis, drug development, and understanding cellular mechanisms. It involves identifying and locating cells within images acquired from various microscopy techniques. In order to understand cell behavior and tissue structure, using computer-aided system is a efficient and promising way. In this paper, we present our approach for the OCELOT 2023 Cell Detection from Cell-Tissue Interaction challenge. Our proposed method utilizes the FC-HarDNet architecture for cell detection and tissue segmentation, which has shown promising results in various computer vision tasks. Additionally, we incorporate tissue segmentation results to aid in the classification of detected cells, leveraging the valuable information encoded in the spatial relationships between cells and their surrounding tissue. Our method achieved **0.6992 mean F1-score** and ranked fifth in the OCELOT 2023 Challenge, demonstrating the potential of integrating cell-tissue interactions for improved cell detection in biomedical image analysis.

Keywords: cell detection · tissue segmentation · FC-HarDNet · cell-tissue interaction

1 Introduction

In the field of histopathology image analysis, cell detection is considered a pivotal task [1,3,5]. It enables the quantification and analysis of different types of cells, offering the potential to enhance disease prognosis assessment and formulate more precise patient treatment plans. Moreover, this task adheres to the principles of medical interpretability, ensuring that medical professionals can comprehend and trust its results.

Given that the outcomes of cell detection can directly impact patients' lives and health, the importance of high-performance cell detection models in practical clinical applications is self-evident. Consequently, ongoing in-depth research and

---

Y.-W. Lo and C.-H. Yang—Contributed equally to this work.

© The Author(s), under exclusive license to Springer Nature Switzerland AG 2024
S.-A. Ahmadi and S. Pereira (Eds.): MICCAI 2023, LNCS 14373, pp. 130–137, 2024.
https://doi.org/10.1007/978-3-031-55088-1_12

improvements are essential to ensure that this task can realize its maximum potential in real-world applications, providing continuous support for patients and healthcare.

In the paper OCELOT: Overlapped Cell on Tissue Dataset for Histopathology [4], the authors propose a novel concept of cell detection using the relationship between cells and tissues. Typical cell detection methods usually infer the location and type of cells by observing their morphology alone, but fail when the morphological features of cells are difficult to categorize individually. However, by introducing the aid of tissue information, the new method offers significant advantages over conventional methods. This innovative approach enables a significant improvement in the accuracy and stability of cell detection in situations with overlapping cells and complex backgrounds. In addition, the method is expected to play a key role in other medical fields and biological image processing tasks, providing strong support for more accurate disease diagnosis and treatment.

Therefore, based on FC-HarDNet for semantic segmentation, we use it as the main model and design a two-branch architecture. One branch deals with the segmentation of tissue pictures and the other branch deals with the prediction of cell locations. Ultimately, we use the segmentation results of the tissue pictures to classify the cells in order to achieve the prediction of background cells and tumor cells.

The contributions of this study can be summarized as follows. We have successfully applied FC-HarDNet to the cell detection task; secondly, we have utilized tissue information to aid cell detection; thirdly, we have evaluated the proposed method using the OCELOT 2023 Challenge.

2   Method

Following a similar approach as presented in [4], we consider cell detection as segmentation task. To effectively address the cell-tissue pairs, we construct a dual-branch framework featuring distinct networks for cell and tissue tasks. In this framework, we employ cell branch to accurately predict cell coordinates, and utilize tissue branch to predict the cell's corresponding class. The architecture of two branches are both FC-HarDNet, which ensures a robust and efficient computational backbone for cell and tissue branch.

2.1   Preliminary

In the cell branch, the cell labels of segmentation task are generated by drawing a fixed-radius circles centered on each cell point annotation, and then filled with the corresponding class label. The cell model is trained based on the generated labels and is responsible for predicting the positions of all cells in the cell picture. On the other hand, the tissue branch is responsible for predicting the categories of different areas of the provided tissue image. During the inference phase, we identify each local peaks within the predicted cell probability maps, then extract

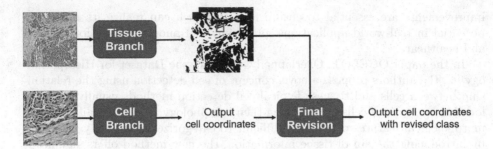

Fig. 1. A dual-branch framework for cell localization and classification tasks.
The tissue branch processes tissue image to predict tissue categories, while the cell
branch handles cell images to predict the coordinates of cells. Subsequently, the predicted cell positions from the cell branch are mapped to the corresponding positions
in the tissue images. The cell types are determined based on the predicted categories
from the tissue branch.

these peaks and present them as cell coordinate predictions. Once we have identified the predicted cell coordinates, we map the predicted cell positions to the
corresponding positions in the tissue image, and determine the type of cell based
on the predicted category of the tissue branch. For example, if a cell is located
in the background area, it is classified as a background cell, and if it is in the
cancer area, it is classified as a tumor cell.

The overall approach is depicted in Fig. 1. This approach works toward
strengthening the cell branch's ability to detect cells, and enhancing the tissue branch's classification capability. Through this dual enhancement strategy,
we optimize the overall performance of both branches, ensuring accurate cell
detection and precise classification within the framework.

2.2   FC-HarDNet

FC-HarDNet (Fully Convolutional HarDNet) is an advanced convolutional neural network architecture designed primarily for semantic segmentation tasks.
Figure 2 shows the intricate structure of FC-HarDNet in cell branch. It consists
of basic building blocks called HarDBlock, which is derived from HarDNet [2],
and is designed to reduce the DRAM access. The unique balance of HarDBlock
between efficiency and performance sets it apart from other models.

2.3   Model Ensemble

To enhance the accuracy of our predictions, we employ an ensemble strategy
that incorporates 5-fold cross-validation and Test Time Augmentation (TTA).
The dataset is randomly divided into five separate folds, each containing 80
images. For each of the five iterations, four folds are used for training and one
for validation, and that results in five distinct sub-models.

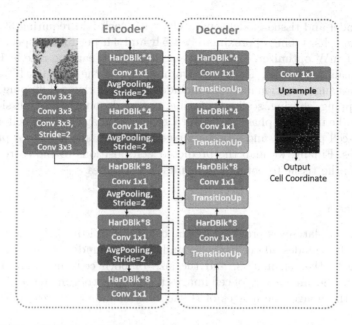

Fig. 2. FC-HarDNet architecture in Cell Branch. The encoder and decoder are composed of "HarDBlk." In the cell branch, when a cell image is input into the model, the output of the model will be a predicted map of cell positions in that image.

During the inference stage, we apply Test Time Augmentation by augmenting each test image with image flipping, creating an additional image. We then feed both the original test images and the augmented ones to the sub-models. The sub-models provide individual predictions for each image. To arrive at the final prediction, we take the average of the outputs generated by the sub-models for each image.

2.4 Loss Function

The loss functions of both cell branch and tissue branch are the same, and given in Eq. (1). It calculates the loss between the ground truth G, the output O of our model.

$$L = l_{BCE}^{w}(G,O) + l_{IoU}^{w}(G,O) \tag{1}$$

where l_{BCE}^{w} and l_{IoU}^{w} denote weighted BCE loss and weighted IoU loss, respectively. These two functions have the same definition as that of [6].

3 Experiment

3.1 Setting

We train our proposed methods on a single NVIDIA Tesla V100 GPU. As mentioned before, both tissue and cell branch are based on FC-HarDNet70. We train the cell model and tissue model separately, with each fold requiring 300 epochs and 1000 epochs, taking approximately 1.5 h and 4 h per fold, respectively. We use the AdamW optimizer. The batch size is 16 and learning rate is 1e-4 with cosine annealing schedule. The image training size of the model is 1024×1024.

Our data augmentation methods include random horizontal flipping, random vertical flipping, shifting, scaling, rotating, coarse dropout, and Gaussian noise.

During the inference phase, we input a given cell image into the cell model for predicting cell positions, and a tissue image into the tissue model for predicting tissue types. Finally, we map the predicted cell positions to the corresponding locations in the tissue to obtain the results of cell types.

3.2   Dataset

The OCELOT dataset is provided by the organizer of MICCAI 2023 OCELOT Challenge. It includes 400 cell images and 400 surrounding tissue images corresponding to the cell images, and the size of both cell and tissue images is 1024×1024. The annotations of cell images include the coordinates and cell types of cells in the image; the images of tissue cells include 3 classes of tissue type annotations.

3.3   Results

As shown in Table 1, we conduct the OCELOT 2023 Cell Detection from Cell-Tissue Interaction challenge using five distinct settings to explore the impact of different training strategies on cell and tissue detection performance. Below, we present the details, results and analysis for each setting.

In Setting 1, we train the cell model using 320 cell images. The cell model is trained to classify each cell image as either background or cell (3 class: background, cell, none). And we use 400 tissue images to train the tissue model, classifying each pixel into its corresponding tissue category. Setting 1 is used as a baseline for comparison with various internal experiments.

Setting 2 focuses on Test Time Augmentation (TTA) for both the cell and tissue models. In Setting 2, we use 400 cell images to train and applied TTA on the cell model to enhance cell detection performance. In the tissue part, TTA is added on the basis of the tissue of Setting 1 to enhance tissue segmentation.

Setting 3 utilizes 5-fold cross-validation ensemble and TTA for the cell model. In Setting 3, we train a cell model using 400 cell images with 5-fold cross-validation ensemble and TTA. The setting of tissue model is the same as Setting 2.

Setting 4 follows a similar approach to Setting 3, incorporating 5-fold cross-validation ensemble and TTA for both the cell and tissue models. The key difference between Setting 3 and Setting 4 lies in the classification granularity of the cell model. Setting 3's cell model has three classes (background, cell, none), while Setting 4's cell model has only a single class (cell).

In Setting 5, we trained the cell model using 400 cell images without using any tissue images in the training process. The cell detection model was solely trained based on cell labels to classify background, background cell, and tumor cell. This setup serves as a comparison to assess the impact of not utilizing tissue images on cell detection.

Table 1. Difference between each experiment setting. The detailed configurations of various experiments involve adjustments in the number of images, class categories in the models, and the inclusion of 5-fold ensemble and test time augmentation. These variations aim to compare the effects of different approaches.

Model-Setting	Cell Branch				Tissue Branch			
	number of image	class	5fold	TTA	number of image	class	5fold	TTA
Setting 1	320	3			400	3		
Setting 2	400	3		✓	400	3		✓
Setting 3	400	3	✓	✓	400	3		✓
Setting 4	400	1	✓	✓	400	3	✓	✓
Setting 5	400	3			–	–	–	–

Table 2. Results of using different experiment setting in the validation set. Compare the performance of different experimental settings on cell detection in the validation set. (the setting details are in Table 1 and the highest mF1-Score is written in bold)

Model-Setting	Background Cell			Tumor Cell			mF1-Score
	Precision	Recall	F1-Score	Precision	Recall	F1-Score	
Setting 1	0.6010	0.6806	0.6383	0.7737	0.7884	0.7810	0.70965
Setting 2	0.5807	0.7052	0.6370	0.7394	0.8119	0.7740	0.70550
Setting 3	0.6128	0.6930	0.6504	0.7828	0.7891	0.7860	**0.71820**
Setting 4	0.6107	0.6842	0.6453	0.8158	0.7585	0.7861	0.71570
Setting 5	0.5425	0.7089	0.6147	0.7742	0.7410	0.7573	0.68600

Our experimental results in the validation set shown in Table 2 demonstrate the impact of different training strategies on cell and tissue detection performance. Setting 1's results provide a baseline for individual model training, while Setting 2 highlights the benefits of TTA in improving detection results and shows that training a model on all training images may lead to worse model training. In Settings 3 and 4, the utilization of 5-fold cross-validation and TTA for model ensembling contributes to more robust and competitive cell and tissue detection. Finally, setting 5's result indicates that removing the tissue branch significantly reduces the predictive capability of our approach.

Considering the benefits of each setting mentioned above, such as the robustness and stability, we choose setting 3 as our final submitted version. In the test set, our approach performs impressively, achieving a notable mean F1-Score of

Table 3. Result of using setting 3 in the test set. (the setting details are in Table 1 and the result of using setting 3 in the validation set is in Table 2)

Background Cell			Tumor Cell			
Precision	Recall	F1-Score	Precision	Recall	F1-Score	mF1-Score
0.6633	0.6424	0.6527	0.6997	0.7980	0.7457	0.6992

0.6992. This accomplishment led to our team securing the fifth position among all participating teams. The overall result of test set is shown in Table 3.

4   Discussion

To confirm the feasibility of our dual-branch architecture, we conducted a statistical analysis of the number of cells belonging to a specific class within the tissue. Table 4 presents the statistical results of the OCELOT dataset. Notably, in the training dataset, 15.4% of background cells are classified as cancer area in the tissue images, while 6.6% of tumor cells are classified as background area. Similarly, this phenomenon was observed in both the validation dataset and the test dataset. Such inconsistencies in labeling on both sides may lead to misclassification in our dual-branch architecture because our strategy involves classifying all cells located in the background area as background cells and all cells in the cancer area as tumor cells. This approach results in the inability to accurately distinguish tumor cells in the background area and background cells in the cancer area. While our dual-branch architecture has some drawbacks in cell classification, sacrificing accuracy in some cell types, it still achieves a comparable accuracy on the OCELOT dataset.

Table 4. The numbers of different cell's types in different tissue regions. This table counts the number of various cell types in different tissue regions in the training, validation, and test sets.

Phase	Type	Background Area	Cancer Area	Unknown Area	Total
Training	Background cell	18951(81.2%)	**3605(15.4%)**	776(3.3%)	23332
	Tumor cell	**2808(6.6%)**	39180(92.1%)	517(1.2%)	42505
Validation	Background cell	6740(80.2%)	**1151(13.7%)**	509(6%)	8400
	Tumor cell	**785(4.8%)**	15317(93.9%)	210(1.2%)	16312
Testing	Background cell	7749(80.6%)	**1549(16.1%)**	305(3.1%)	9603
	Tumor cell	**699(5.4%)**	11985(93%)	190(1.4%)	12874

5   Conclusion

Our study introduces a novel approach to the OCELOT 2023 Cell Detection from Cell-Tissue Interaction challenge, leveraging the power of the FC-HarDNet architecture for cell detection and tissue segmentation. By effectively incorporating tissue segmentation results, we demonstrate the ability to leverage the spatial relationship between cells and their neighboring tissues to improve the accuracy of cell classification. However, it is important to acknowledge a limitation in our method related to cell class assignment. Specifically, our approach tends to classify all cells in the cancer area as tumor cells and all cells in the background area as background cells. This oversimplified class assignment does not fully capture the heterogeneity of cell types in both cancer and background areas. Despite this limitation, during the testing phase of the OCELOT 2023 Challenge, our method achieved a competitive mean F1-score of 0.6992 and ranked fifth among all participating teams.

Acknowledgments. This research is partially supported by the Ministry of Science and Technology (MOST) of Taiwan. We thank the National Center for High-performance Computing (NCHC) for computational and storage resources.

References

1. Al-Kofahi, Y., Lassoued, W., Lee, W., Roysam, B.: Improved automatic detection and segmentation of cell nuclei in histopathology images. IEEE Trans. Biomed. Eng. **57**(4), 841–852 (2010)
2. Chao, P., Kao, C.Y., Ruan, Y.S., Huang, C.H., Lin, Y.L.: HarDNet: a low memory traffic network. In: Proceedings of the IEEE/CVF International Conference on Computer Vision, pp. 3552–3561 (2019)
3. Pal, A., et al.: Deep multiple-instance learning for abnormal cell detection in cervical histopathology images. Comput. Biol. Med. **138**, 104890 (2021). https://www.sciencedirect.com/science/article/pii/S0010482521006843
4. Ryu, J., et al.: OCELOT: overlapped cell on tissue dataset for histopathology. In: Proceedings of the IEEE/CVF Conference on Computer Vision and Pattern Recognition, pp. 23902–23912 (2023)
5. Su, H., Xing, F., Kong, X., Xie, Y., Zhang, S., Yang, L.: Robust cell detection and segmentation in histopathological images using sparse reconstruction and stacked denoising autoencoders. Med. Image Comput. Comput. Assist. Interv. **9351**, 383–390 (2015). https://api.semanticscholar.org/CorpusID:9524314, MICCAI International Conference on Medical Image Computing and Computer-Assisted Intervention
6. Wei, J., Wang, S., Huang, Q.: F^{3}net: fusion, feedback and focus for salient object detection. In: Proceedings of the AAAI Conference on Artificial Intelligence (2020)

Dense Prediction of Cell Centroids Using Tissue Context and Cell Refinement

Joshua Millward[✉][iD], Zhen He[iD], and Aiden Nibali[iD]

School of Computing, Engineering and Mathematical Sciences, La Trobe University,
Victoria, Australia
j.millward@latrobe.edu.au

Abstract. Cell detection is a common task in computational pathology, often fundamental for downstream tasks that can aid in predicting prognosis or treatment response. The Overlapped Cell on Tissue Dataset for Histopathology (OCELOT) challenge aimed to explore ways to improve automated cell detection algorithms by leveraging surrounding tissue information. We developed two cell detection algorithms for this challenge that both leverage surrounding tissue context to enhance their performance. The first is fed an additional input representing a cancer area probability heatmap, predicted from a tissue segmentation model. The second is fed the cancer area probability heatmap, in addition to a heatmap representing cell locations predicted from a separate model. Submitting our first algorithm, we achieved a mean F1 score of 74.73 on the challenge validation set, and second place with a mean F1 score of 72.21 on the challenge test set. Our algorithms do not require paired cell and tissue annotations to train, enabling their use to enhance existing cell detection models where paired annotations may not exist.

Keywords: Cell Detection · Segmentation · Deep Learning

1 Introduction

Detecting the presence of particular cell types in tissue samples is a task commonly performed by pathologists to aid in the prediction of patient outcome or likely response to therapy. Examples of this include the assessment of tumour-infiltrating lymphocytes [9], and mitoses [4]. Manually making these assessments is time consuming, and subject to inter- and intra-rater variability, highlighting automated tools as a solution to address these limitations.

The advent of computational pathology has enabled the automated assessment of whole slide images (WSI). Deep learning based methods are popular in computational pathology, with approaches ranging from black-box methods that generate a single classification for a slide [2], to interpretable methods that detect cells or tissue structures within a slide [6].

When manually quantifying cells, pathologists will often start by analysing the sample at a low magnification to better understand broad tissue structure,

S.-A. Ahmadi and S. Pereira (Eds.): MICCAI 2023, LNCS 14373, pp. 138–149, 2024.
https://doi.org/10.1007/978-3-031-55088-1_13

then use a higher magnification to inspect the tissue microenvironment and identify individual cells. The OCELOT challenge [8] was developed to explore whether automated cell detection algorithms could be enhanced by following a similar methodology in incorporating cell-tissue relationships. Participants of the challenge were required to submit an algorithm to detect tumour cells (TC) and background cells (BC) in a dataset released alongside the challenge, and were evaluated on the mean F1 score across regions in a held-out test set.

In addition to the challenge, the OCELOT organisers implemented their own algorithms that consider cell-tissue relationships, noting that there were no existing studies that had attempted this [8]. Their most successful approach on the OCELOT validation set, *pred-to-input*, involved feeding in predicted tissue probabilities from an auxiliary tissue segmentation branch in their network to the input of their cell detection branch.

In this work we present two of our approaches to cell detection that leverage cell-tissue relationships. Our first, *tissue context* algorithm follows a similar approach to the *pred-to-input* OCELOT approach [8], however has some differences. Firstly, we take a different approach to cell detection, framing it as a dense prediction task. Secondly, we use a transformer-based model architecture for our segmentation and detection models, which we believe better make use of contextual information than traditional CNN-based models [1]. Thirdly, we tile the input data at inference time in a way that overcomes boundary effects. Finally, we identify a set of under-annotated examples in the training dataset which we exclude from model development.

Our second, *cell refinement* algorithm has similarities with our first, except employs three separate neural networks. The first is the same tissue segmentation model used in our tissue context algorithm. The second is a cell detection model that outputs a heatmap of cell locations that is agnostic to cell class using only the image data. The third network is a cell detection model that ingests the image data, along with heatmaps from the tissue and cell models, and generates a set of per-class cell detections. This allows the third network to act as a 'refinement' model, that can focus on detecting and classifying cells, and 'fixing' any incorrectly predicted cell locations. This approach advocates a separation of concerns where one model is used to detect any kind of cell and a separate model to classify and refine cell detections.

We submitted our tissue context algorithm to the challenge test set and achieved second place on the leaderboard, showing the success of our relatively simple approach over others that used more complex techniques such as ensembling.

We briefly summarise the contributions from this work below:

1. We frame cell detection as a dense prediction task by rendering target heatmaps from ground truth cell point annotations.
2. We separately train transformer-based models for tissue segmentation and cell detection, feeding the cell detection model a heatmap representing predicted cancer areas from the tissue segmentation model.

3. We propose a cell refinement algorithm that takes a predicted cell heatmap as input and further refines the heatmap using both the image data and the output of a tissue segmentation model.
4. We show that with our approach, existing cell detection models can leverage tissue information without needing paired tissue segmentation and cell detection annotations to exist within the same dataset.
5. We make our training and inference pipeline, including model weights submitted to the OCELOT challenge, publicly available on GitHub: https://github.com/Mward94/ocelot-segdet.

2   Methodology

2.1   Data

There were 400 'tissue' and 'cell' training image pairs released with the OCELOT challenge. Images were provided as 1024×1024 pixel crops, with tissue images at lower magnification (≈ 0.8 microns-per-pixel (MPP)), than cell images (≈ 0.2 MPP).

Tissue image annotations were provided as segmentation masks with three classes annotated: background, cancer area, and unknown. Point annotations were made within cell images capturing tumour cells and background/other cells.

We initially split the images into an internal training and validation set with a 90%/10% distribution to enable quantification of model performance. Following the creation of an initial cell detection model, analysis of performance on the internal validation set showed a high reported number of false positive detections for a small number of images. Upon manual inspection of these examples, it was found that a large number of cells being detected by the model were absent from the ground truth annotations despite visual evidence for their existence.

To reduce the negative impact on model performance due to potentially under-annotated images, a visual inspection of all cell annotations was performed, where 40 cell images were identified as having varying degrees of visually identifiable cells that were not annotated, and 3 cell images were identified as having no tissue content. To avoid training on poor quality data it was decided to exclude the 43 identified cell images from cell detection model development (creating a 'refined training set'), whilst using all 400 tissue images for tissue segmentation model development. To facilitate this, two internal training/validation splits were created to measure performance during model development. A 90%/10% (321/36) split for the cell images, and a 91%/9% (364/36) split for the tissue images, such that both validation sets contained the same pairs of cell/tissue images. Image IDs of the excluded cell images are listed in Appendix A.

2.2   Tissue Context Algorithm

Figure 1 illustrates our proposed tissue context algorithm, consisting of two separate, independently trained, deep learning models. The first model performs tissue segmentation (cancer area vs. normal tissue & background), and generates a

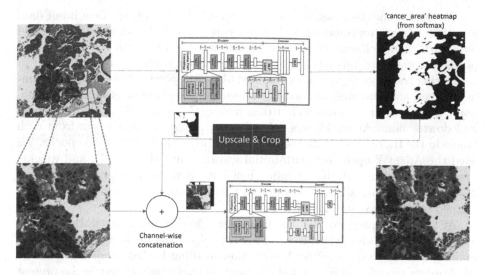

Fig. 1. Proposed *tissue context* cell detection algorithm. This consists of separately trained models for tissue segmentation and cell detection. The cell detection model is fed the cell image and softmaxed 'cancer area' prediction heatmap from the tissue model. Both models use the SegFormer [10] architecture, however any semantic segmentation architecture could be substituted in their place.

probability heatmap per-pixel, per-class. We extract the 'cancer area' heatmap, then upscale and crop it to get the equivalent spatial area represented by the cell image. The heatmap is concatenated with the cell image along the channel dimension, then fed to the second model to perform cell detection (tumour cells and background cells). The tissue segmentation model is trained with data at a lower magnification than the cell detection model, effectively providing the cell detection model more context than it would otherwise have.

All models were trained using the open-source PyTorch [7] package. Further detail on each model is provided in the following sections.

Tissue Segmentation. We trained a SegFormer-B0 model [10] on the tissue images, beginning with initial backbone weights pretrained on ImageNet. A driver for choosing this architecture is that we believe the self-attention mechanisms inherent in transformer-based models can more effectively leverage contextual information present in the images.

Image data was normalised using the Macenko [5] stain normalisation technique, with tissue masking applied to ignore background pixels from the normalisation process. The masking procedure involved converting the RGB image to the LAB colour space, then retaining pixels with a luminosity value < 0.8.

Images were fed to the model in 512×512 pixel patches, with image data scaled to 0.5MPP, equivalent to $256\,\mu m^2$ of contextual information.

We trained the tissue segmentation model to semantically segment input data into three classes, representing each of the 'cancer area', 'background tissue', and 'unknown tissue' classes. Cross entropy loss was used to train the model, with pixels annotated as 'unknown tissue' excluded from gradient calculations.

During training, we randomly sampled 512×512 pixel regions from the training images, and randomly applied the following suite of augmentations: rotation, scale, horizontal flips, Gaussian blurring, adding Gaussian noise, colour jittering, and downscaling. An epoch was defined as two random crops taken from each image in the training set, with the final model being trained for 1500 epochs. We used the AdamW optimiser, with initial learning rate of 6×10^{-5}, and weights linearly decayed to 0 during training. Following hyperparameter tuning on the 91%/9% data split, all data was moved to the training set and a final model trained on all 400 tissue images.

At inference time, we 'tile' the image, and feed it to the network in 512×512 pixel tiles. This allows us to apply our algorithm to images of any size, making it fit for use on WSIs. A limitation to tiling is that predictions on the tile borders are missing contextual information that would otherwise be present in the full image. To overcome this, we define an 'output crop margin' per-tile, where predictions are only retained within some margin of the tile border. For this model we use an output crop margin of 64 pixels (illustrated in Appendix B). To ensure predictions are generated for the entire image, tiles are extracted at 512×512 pixels, but strided by the retained area. Predictions are generated for all tiles, then combined to form a single prediction map (if multiple tiles contain predictions for the same area, predictions from the bottom-right most tile are used).

We apply a softmax to the reconstructed prediction map to generate per-pixel probabilities per-class, then extract the heatmap corresponding to the 'cancer area' to feed to the cell detection model.

Cell Detection. We approach cell detection as a dense prediction task instead of following traditional object detection methods for two reasons. Firstly, annotations were provided as points, meaning in order to train a typical object detector, we would need to estimate cell boundaries, which could be highly error prone due to the varying size and shape of cells within a single class (particularly background cells). Secondly, we again anticipated that the self-attention mechanisms inherent in transformer-based models could more effectively leverage the contextual information required to make accurate predictions. As such, using the cell images, we trained a dense prediction cell detection model using a SegFormer-B2 model, with initial backbone weights pretrained on ImageNet.

To leverage tissue information in the cell detection model, we concatenate the input image data with a corresponding predicted cancer area heatmap. By using predicted cancer area heatmaps generated by the tissue segmentation model instead of ground truth tissue annotations, we eliminate the need for paired cell-tissue annotations. We initially experimented with using an argmaxed heatmap, however found better performance by providing the softmaxed heatmap, which we believe is due to additional information in the soft assignment as opposed to the hard assignment. Data was fed to the model in 512 × 512 pixel patches, with image data (and cancer area heatmap) scaled to 0.2MPP, equivalent to $102.4 \mu m^2$ of contextual information in a patch.

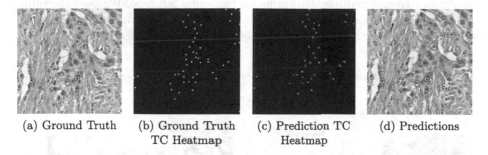

(a) Ground Truth   (b) Ground Truth   (c) Prediction TC   (d) Predictions
                         TC Heatmap   Heatmap

Fig. 2. Examples of generated ground truth and predicted heatmaps for the TC class when performing cell detection via dense prediction. In the overlays, blue corresponds to tumour cells, whilst yellow corresponds to background cells. (Color figure online)

We convert the ground truth point annotations to a target heatmap by placing a 2D Gaussian distribution centred on each ground truth coordinate, with separate channels corresponding to each cell class (in our case, no cells, background cells, and tumour cells). The standard deviation of the Gaussian was chosen to be approximately $1.14 \mu m$, such that there was an $8 \mu m$ diameter within 7 standard deviations. This size was chosen such that the Gaussian was contained within the bounds of most cells. Values in the heatmap were ensured to be in the range: $[0, 1]$. An example of the ground truth point annotations and generated target heatmap for the TC class can be seen in Figs. 2a and 2b.

To train the model we apply the Sigmoid function to the model output, then compute the Mean Square Error (MSE) loss between the output and ground truth heatmap (ignoring the channel corresponding to no cells). We found better performance when not applying Macenko stain normalisation, which we attribute to poor results arising from images containing small amounts of tissue.

Following hyperparameter tuning on the 90%/10% data split, all data was moved to the training set and a final model trained on the refined 357 cell images. The same set of augmentations were applied as for tissue segmentation, except for adding Gaussian noise. An epoch was defined as one random crop taken from each image in the training set, with the final model being trained for 250 epochs. The AdamW optimiser was used, with initial learning rate of 6×10^{-5}, and weights linearly decayed to 0 during training.

Following training, the model outputs a heatmap where per-channel, each pixel corresponds to how likely it belongs to a cell of that class. These heatmaps consist of circular, blob-like areas with varying degrees of likelihood. To convert this into a set of per-class point coordinates, we apply a blob detection algorithm using OpenCV in each channel of the heatmap. During experimentation we found better blob extraction by first applying Gaussian modulation to the heatmap (with same standard deviation as for ground truth generation). The blob detection algorithm can reject detection of cells in areas of low likelihood, as seen in Figs. 2c and 2d, where faint blobs are not extracted as tumour cells. After extracting detections we apply a point-based non-maximum suppression algorithm to avoid multiple detections for the same cell type. We used the same tiling procedure as for tissue segmentation, with a 128 pixel output crop margin. Model outputs were first reconstructed before cell coordinates extracted.

2.3   Cell Refinement Algorithm

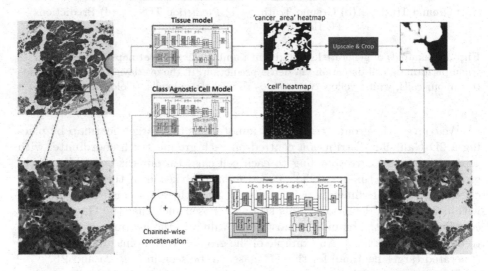

Fig. 3. Proposed *cell refinement* algorithm. Separate tissue segmentation and class agnostic cell detection models are trained that are only fed image data as input. The cancer area heatmap is extracted from the tissue segmentation model, and the cell heatmap from the class agnostic cell detection model. The heatmaps are concatenated with the cell image, before being fed to a third SegFormer [10] model.

Figure 3 illustrates our proposed cell refinement algorithm, consisting of three separate, independently trained models. The first is the same tissue segmentation model as described in our tissue context algorithm. The second is a class agnostic cell detection model that is only fed image data, and generates a class agnostic cell heatmap. The third model is similar to our tissue context cell detection model, except also has the cell heatmap concatenated to its input.

The intuition behind this approach was that the first two models would generate a set of initial predictions on tissue and cell locations, allowing the third model to focus on detecting and classifying cells, whilst 'refining' potential incorrectly predicted cell locations. We found the cell refinement model performance better when a class agnostic cell heatmap was used instead of separate heatmaps per class. We think the reason for this is that not providing the class information forces the cell refinement model to work harder to determine the class information, which produces more useful feature maps. This is similar to the benefits of masking the input for the masked autoencoder [3]. Practically, this approach enables a general cell detector to be trained across multiple existing datasets, and be reused in different tasks, where only the refinement model would need to be trained for the downstream task.

During model development, we observed that the refinement model was over-fitting to the training set, which we believed was due to the incoming cell heatmaps being mostly correct, resulting in the refinement model not learning to refine the predictions.

To address overfitting, we created a database of over 10,000 blobs extracted from various heatmaps, and implemented a 'blob addition' augmentation, where during training we would randomly sample up to 10 blobs from our database and add them to random locations in the incoming cell heatmap, forcing the refinement model to correct these incorrect cell locations.

3   Results

Table 1. Results on the OCELOT [8] challenge validation set for different algorithms. Refined dataset refers to cell models that were trained on the refined 357 cell images. The **highest score** in each section is written in bold.

Algorithm	Refined Dataset	mF1	F1 TC	F1 BC
OCELOT Cell-only [8]	–	68.87	–	–
Ours Cell-only	–	70.25	78.12	62.38
Ours Cell-only	✓	**70.89**	77.73	64.04
OCELOT pred-to-input [8]	–	73.36	–	–
Tissue Context	–	73.45	80.45	66.44
Cell Refinement	–	72.84	79.72	65.95
Cell Refinement + Blob Addition	–	**74.33**	81.30	67.36
Tissue Context	✓	**74.73**	80.73	68.72
Cell Refinement	✓	73.79	79.84	67.73
Cell Refinement + Blob Addition	✓	74.12	80.34	67.89

Our tissue context algorithm achieved second place on the OCELOT test set leaderboard with a mean F1 score of 72.21.

To further compare the performance of our algorithms and to contrast them to the original OCELOT results [8], we performed an evaluation over the challenge validation data that was released after the challenge concluded. These results are presented in Table 1.

To assess the value added by feeding additional information to the cell detection model, we first evaluate the performance of a 'cell-only' version of our algorithm, which is a modified version of our tissue context cell detection model that is only fed image data at input. We observe that our model outperforms the equivalent OCELOT model by an absolute 1.38 in terms of mean F1 score, and also see a slight improvement in the same model when trained on the refined dataset.

When evaluating our models trained on the entire dataset, we see our tissue context algorithm has similar performance to the equivalent OCELOT algorithm, however our cell refinement algorithm trained with the blob addition augmentation outperformed the OCELOT result by an absolute 0.97 in terms of mean F1 score. The blob addition augmentation appeared to be key in achieving improved performance and helping to address model overfitting.

Our tissue context algorithm trained on the refined dataset achieved the best overall mean F1 score, outperforming the OCELOT result by an absolute 1.37. Our cell refinement algorithm showed improved performance when trained on the refined dataset, though slightly worse performance when blob addition augmentations were also used. We plan to inspect this further in the future.

4   Discussion and Conclusion

In this work we presented two algorithms for cell detection that leverage cell-tissue relationships for enhanced performance.

Within each algorithm, we train our models independently and at different scales, with tissue segmentation models fed data at lower magnification than the cell detection models. This allows the tissue segmentation models to see a wider context of the cancer, whilst allowing the cell detection models to see a detailed view of the tissue microenvironment, more closely mimicking the workflow of a pathologist. By feeding the cancer area probability heatmap to the cell detector, we effectively allow it to use information that it would otherwise be lacking context to predict itself.

When evaluating a cell-only variant of our approach, we saw a large improvement in terms of mean F1 performance over the equivalent OCELOT model, suggesting that our approach to cell detection as a dense prediction task, our selection of model, and our approach to tiling are all highly beneficial choices to create a good cell detection algorithm.

We identified that training with the refined cell dataset greatly improved cell detection performance of our tissue context algorithm, primarily due to a boost in background cell performance, highlighting the impact that under-annotated training data can have on model performance.

Our cell refinement algorithm with blob addition augmentations achieved very good performance when trained on the whole dataset, but saw no notable improvement when trained on the refined dataset. It is possible the impact of the augmentation is limited by less overall data variation in the refined dataset. We intend to continue working on this algorithm in the future by applying it to multiple, larger datasets, and exploring other types of augmentations such as the random removal of blobs from the incoming cell heatmaps.

To explore whether further improvements could be made to model performance, we have identified two other possible directions for future work. First, we are interested in exploring whether performance can be enhanced by ensembling predictions from multiple models, and particularly what the trade-off would be in terms of model inference time. Second, we are interested in comparing the performance of other semantic segmentation architectures (including CNN-based ones) for our segmentation and dense prediction models, to see if better performance could be achieved.

An existing limitation in most cell detection models is the inability to mimic the way pathologists manually detect cells. Namely, utilising cell-tissue relationships to detect and classify cells. Approaches that take paired cell and tissue image annotations as input to model training require specialised paired annotations. In contrast, a major strength of both of our algorithms is that we can still leverage benefits of cell detection using tissue segmentation information, without paired cell and tissue annotations, due to our models being trained in isolation. This means our approach can be trained on widely available detection or segmentation datasets, and still leverage the benefits of detecting cells using tissue information.

The competitive performance achieved by our algorithms demonstrates their potential to improve cell detection, and by publicly releasing our code, we hope the research community may build on these ideas to further improve performance.

A   Excluded Image IDs

The following 'cell' image IDs from the released training set were identified as containing varying degrees of under-annotated cells and excluded from cell detection model development. The corresponding 'tissue' images were still used to train the tissue segmentation model.

IDs: 051, 074, 079, 129, 135, 138, 140, 144, 147, 152, 168, 172, 181, 201, 223, 233, 244, 249, 251, 252, 255, 256, 263, 267, 279, 286, 292, 294, 307, 315, 323, 325, 334, 341, 345, 352, 376, 393, 396, 397.

Macenko normalisation on the 'cell' images failed for the following image IDs due to containing no tissue. These were excluded from cell detection model development. The corresponding 'tissue' images were still used to train the tissue segmentation model.

IDs: 042, 217, 392.

B   Output Crop Margin

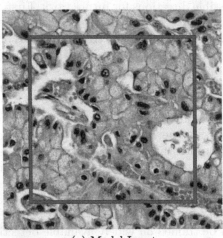

(a) Model Input   (b) Model Predictions

Fig. 4. Illustration of the output crop margin applied to the tissue segmentation model. Given a 512×512 pixel input tile (a), predictions are retained for the inner 384×384 pixels (red box), within 64 pixels of the border (b). Red overlay corresponds to predicted cancer area, whilst blue corresponds to normal tissue or background. (Color figure online)

References

1. Atabansi, C.C., Nie, J., Liu, H., Song, Q., Yan, L., Zhou, X.: A survey of transformer applications for histopathological image analysis: new developments and future directions. Biomed. Eng. Online **22**(1), 96 (2023)
2. Chen, R.J., et al.: Scaling vision transformers to gigapixel images via hierarchical self-supervised learning. In: Proceedings of the IEEE/CVF Conference on Computer Vision and Pattern Recognition, pp. 16144–16155 (2022)
3. He, K., Chen, X., Xie, S., Li, Y., Dollár, P., Girshick, R.: Masked autoencoders are scalable vision learners. In: Proceedings of the IEEE/CVF Conference on Computer Vision and Pattern Recognition, pp. 16000–16009 (2022)
4. Kadota, K., et al.: A grading system combining architectural features and mitotic count predicts recurrence in stage i lung adenocarcinoma. Mod. Pathol. **25**(8), 1117–1127 (2012)
5. Macenko, M., et al.: A method for normalizing histology slides for quantitative analysis. In: 2009 IEEE International Symposium on Biomedical Imaging: from Nano to Macro, pp. 1107–1110. IEEE (2009)
6. Pai, R.K., et al.: Quantitative pathologic analysis of digitized images of colorectal carcinoma improves prediction of recurrence-free survival. Gastroenterology **163**(6), 1531–1546 (2022)

7. Paszke, A., et al.: PyTorch: an imperative style, high-performance deep learning library. In: Advances in Neural Information Processing Systems, vol. 32, pp. 8024–8035 (2019)
8. Ryu, J., et al.: OCELOT: overlapped cell on tissue dataset for histopathology. In: Proceedings of the IEEE/CVF Conference on Computer Vision and Pattern Recognition (CVPR), pp. 23902–23912 (2023)
9. Williams, D.S., et al.: Lymphocytic response to tumour and deficient DNA mismatch repair identify subtypes of stage ii/iii colorectal cancer associated with patient outcomes. Gut **68**(3), 465–474 (2019). https://doi.org/10.1136/gutjnl-2017-315664
10. Xie, E., Wang, W., Yu, Z., Anandkumar, A., Alvarez, J.M., Luo, P.: SegFormer: simple and efficient design for semantic segmentation with transformers. Adv. Neural. Inf. Process. Syst. **34**, 12077–12090 (2021)

Enhancing Cell Detection in Histopathology Images: A ViT-Based U-Net Approach

Zhaoyang Li[1], Wangkai Li[1], Huayu Mai[1], Tianzhu Zhang[1,2(✉)], and Zhiwei Xiong[1,2]

[1] University of Science and Technology of China, Hefei, China
{lizhaoyang,lwklwk,mai556}@mail.ustc.edu.cn
[2] Institute of Artificial Intelligence, Hefei Comprehensive National Science Center, Hefei, China
{tzzhang,zwxiong}@ustc.edu.cn

Abstract. Cell detection in histology images is a pivotal and fundamental task within the field of computational pathology. Recent advancements have led to the introduction of the OCELOT dataset, which offers annotated images featuring overlapping cell and tissue structures derived from diverse organs. The significance of OCELOT dataset lies in its provision of valuable insights into the intricate relationship between the surrounding tissue structures and individual cells. Based on the OCELOT dataset, We propose a ViT-based U-Net (Cell-Tissue-ViT) in a unified deep model via an encoder-decoder structure for robust cell detection, simultaneously leveraging tissue and cell information. Specifically, we adopt the pretrained ViT encoder of the large-scale pre-trained Segment Anything Model(SAM) as our backbone to enhance the feature extraction capability of the model and adopt LoRA to fine-tune the backbone, intending to enhance its suitability for our specific task. Our approach achieves highly promising results in cell detection on the OCELOT dataset, with an F1-detection score of 0.7558, as indicated by the preliminary results on the validation set. What's more, we achieved **1st** place on the official test set. The code is available in https://github.com/Lzy-dot/OCELOT2023.git.

Keywords: Cell Detection · Computational Pathology · Vision Transformer · Low-Rank Adaptation

1 Introduction

Computational Pathology (CPATH) [1,10,12,16–18,20] is a specialized field within digital pathology that focuses on developing methodologies for the analysis of digitized patient specimens. Among the crucial tasks in CPATH, cell detection in histology images holds vital importance. It facilitates the quantification and analysis of distinct cell types, enabling the extraction of high-level medical insights and leading to improved prognosis evaluation and better patient

Table 1. Cell detection mean **F1** scores of different methods based on our Cell-Tissue-ViTmodel, the **highest score** is highlighted in bold, and the <u>second highest score</u> is underlined.

OCELOT Val	Method		
	Input+Add	Intermediate+Add	Output+Add
F1 Score	<u>0.7442</u>	**0.7518**	0.7208

outcomes. To achieve accurate localization and classification of cells, detailed morphological features such as color and shape are needed. As a result, cell detection datasets are typically collected at high magnification but with a small Field-of-View (FoV).

However, such small FoV datasets inevitably lose the broader context information, which is essential as it provides valuable insights into how cells are arranged and grouped to form higher-level tissue structures. With this premise in mind, the OCELOT dataset [14] was created, containing cell and tissue annotations in small and large FoV patches with overlapping regions, respectively. The question then arises: how can we fully leverage tissue information to enhance the precision of cell detection?

Motivated by [14], we employ a multi-task learning strategy to cope with this problem. Specifically, we utilize a two-branch network to handle the tasks of tissue segmentation and cell segmentation separately and subsequently feed the information from the tissue branch into the cell branch to enhance its perception of the broader context. So, *what kind of tissue information* should we feed into the cell branch, and *in what way*?

After an exhaustive experimental exploration, we find that the most effective strategy, as illustrated in Fig. 2, is to combine the tissue branch's prediction results with the intermediate features from the cell branch encoder via elementwise addition. A comparison of the results from various experiments is presented in Table 1. Besides, to enhance the generalization ability of the model, we utilize the pretrained ViT encoder from the Segment Anything Model (SAM) [9] as our backbone and employ the contemporary model fine-tuning method LoRA [7] for parameter-efficient fine-tuning.

Our main contributions are summarized as follows:

- We propose a ViT-based multi-task learning framework via an encoder-decoder structure for robust cell detection that simultaneously utilizes both cell and tissue information.

- We utilize the pretrained ViT encoder from the extensively pre-trained Segment Anything Model (SAM) as our backbone to enhance the generalization ability of the model. Additionally, we employ the contemporary model fine-tuning method LoRA to fine-tune the backbone, intending to enhance its suitability for our specific task.

- Experimental results on the official validation set demonstrate the effectiveness of our method. What's more, we achieved **1st place** on the official test set!

2 Related Work

2.1 Cell Detection and Segmentation

Cell detection and segmentation stand as crucial pillars in the realm of biological image analysis, catering to a myriad array of experimental methodologies and imaging techniques. The advent of deep learning has heralded a new frontier in the quest for versatile cell segmentation solutions, spurring the development of algorithms with a broad applicability spectrum, as highlighted in some seminal works like [4,11,15]. Pioneering this transformative journey, the U-Net architecture [13] marked the commencement of a groundbreaking epoch in medical image segmentation. In the wake of U-Net, a surge of inventive adaptations have emerged, striving to augment the finesse of cell segmentation. Noteworthy among these are the U-Net++ [24] and 3D-Unet [3], each extending the legacy of their predecessor through enhanced performance and nuanced feature extraction capabilities.

2.2 Large-scale Pre-training Model

Pre-training a Vision Transformer on a large amount of data is an essential process that equips the model with rich, meaningful representations, drawing significant attention to models such as SAM [9] and their diverse applications [21–23]. Despite their outstanding zero-shot generalization abilities, these models often face difficulties in mapping precise anatomical structures without specialized datasets to guide them. Therefore, it is imperative to explore more efficient methods of fine-tuning these large-scale models to adapt to specific downstream tasks, ensuring their applicability and effectiveness in task-specific challenges.

2.3 Parameter-Efficient Fine-Tuning Strategies

Harnessing the power of large-scale pretrained models like SAM [9], which are capable of capturing sophisticated features through expansive data learning, is at the forefront of current research. The challenge lies in effectively adapting these models to impart new, task-specific knowledge to downstream applications. Contemporary approaches primarily focus on refining fine-tuning methodologies to transfer this new knowledge into pretrained behemoths. Rather than overhauling the entire parameter set of these vast models, methods such as visual prompt tuning [8] suggest only adjusting the prompts and the head, thereby delivering high performance with less computational expenditure. Furthermore, LoRA [7] introduces a nuanced strategy, proposing a low-rank approximation to incrementally update parameters within transformer blocks.

3 Method

3.1 Preprocessing

Each sample of the OCELOT dataset $\mathcal{D}$ is composed of six components, $\mathcal{D} = \{(x_s, y_s^c, x_l, y_l^t, c_x, c_y)_i\}_{i=1}^{N}$, where x_s and x_l are the small and large FoV patches extracted from the WSI, y_s^c and y_l^t refer to the corresponding cell and tissue annotations, respectively, and c_x and c_y are the relative coordinates of the center of x_s within x_l. To address the cell-tissue sample pairs, denoted as (x_s, y_s^c) and (x_l, y_l^t), we propose a dual-branch architecture comprising separate networks for cell and tissue segmentation tasks. Drawing inspiration from [19], we formulate cell detection as a segmentation task. In specific, cell labels are represented as segmentation maps, where each cell point annotation is enclosed within a fixed-radius circle centered on the location of the cell and filled with the corresponding class label, as shown in Fig. 1(a). Figure 1(b) illustrates the result of converting the annotation of a cell patch into a semantic segmentation mask. During inference, we identify local peaks within the predicted cell probability maps and consider them point predictions. This enables us to effectively detect and delineate cell structures. Besides, We apply normalization to the images before they are fed into the network. In detail, we scale the intensity level of the RGB images to the range of [0.01, 0.99].

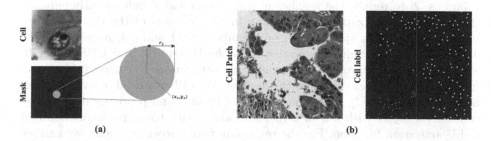

Fig. 1. Example of cell mask generated from the annotation. The left part illustrates the schematic diagram of how a mask is generated for a single cell, where $r_c = 1.4\mu m$, corresponding to 7 pixels at a resolution of 0.2 Microns-per-Pixel (MPP), while the right part shows the corresponding schematic diagram of the entire cell patch generating the semantic segmentation label mask.

3.2 Cell-Tissue-ViT

We developed a novel deep learning network based on a U-Net-shaped encoder-decoder architecture similar to CellViT [6]. Specifically, we employed the pre-trained $SAM - ViT_H$ as our backbone, freezing its parameters and adapting it to our task using LoRA fine-tuning method. Notably, in contrast to [6], we

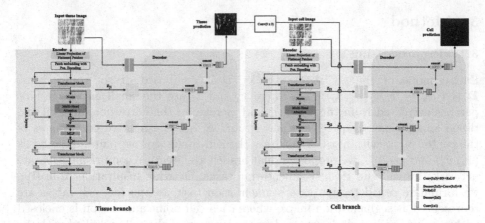

Fig. 2. The Cell-Tissue-ViT network comprises two branches: the tissue branch and the cell branch. Within each branch, a ViT (Vision Transformer) encoder is connected to multiple upsample layers using skip connections. The prediction results from the tissue branch undergo a convolution transformation and are then element-wise added to the multi-level features extracted by the encoder in the cell branch, aiming to provide contextual information that enhances the precision of cell detection.

replaced the convolutional layers in the decoder with depthwise separable convolutions [2] to reduce the number of parameters and simultaneously enhance the model's generalization performance. In the decoders of the tissue and cell branches, dropout is applied with probabilities of 0.1 and 0.3, respectively. As depicted in Fig. 2, taking inspiration from the U-Net [13] and UNETR [5,20] architectures, our approach incorporates skip connections to exploit information from multiple encoder depths in the decoder. We utilize a total of five skip connections. The first skip connection takes the input x and processes it using two convolution layers with a 3 3 kernel size, along with batch normalization and ReLU activation function. For the remaining four skip connections, we extract the intermediate and bottleneck latent tokens z_j, $j \in \{\frac{L}{4}, \frac{2L}{4}, \frac{3L}{4}, L\}$, where L denotes the number of Transformer blocks. Each feature map denoted as z_j, undergoes a series of transformations to increase its resolution by a factor of two. This process involves a combination of deconvolutional layers, which perform upsampling to increase the spatial resolution, and convolutional layers, which modify the latent dimension. Subsequently, the modified feature maps are sequentially processed in the next decoder layer. The prediction results from the tissue branch undergo a convolution transformation and are then element-wise added to the multi-level features extracted by the encoder in the cell branch, aiming to provide contextual information that enhances the precision of cell detection.

3.3   Loss Function

For tissue patch x_t and cell patch x_c with corresponding ground-truth y_t^l and y_s^c, we use the summation of Dice loss and CrossEntropy loss to supervise two branches.

$$\mathcal{L}_{cls}^t = \frac{1}{N} \sum_{i=1}^{N} (\mathcal{L}_{dice}(P_t, y_t^l) + \mathcal{L}_{ce}(P_t, y_t^l)), \tag{1}$$

$$\mathcal{L}_{cls}^c = \frac{1}{N} \sum_{i=1}^{N} (\mathcal{L}_{dice}(P_c, y_s^c) + \mathcal{L}_{ce}(P_c, y_s^c)), \tag{2}$$

where N denotes the number of labeled images, P_t/P_c means the segmentation result of the tissue branch and cell branch, respectively.

4   Implementation Details

4.1   Environment Settings

The development environments and requirements are presented in Table 2.

Table 2. Development environments and requirements.

System	Ubuntu 22.04.1 LTS
CPU	Intel(R) Xeon(R) CPU E5-2695 v4 @ 2.10 GHz
RAM	16×4 GB; 2.67 MT/s
GPU (number and type)	4 NVIDIA Titan RTX (24G)
CUDA version	11.0
Programming language	Python 3.8.5
Deep learning framework	Pytorch (Torch 1.7.1, torchvision 0.8.2)
Specific dependencies	monai 0.9.0
Code	https://github.com/Lzy-dot/OCELOT2023.git

4.2   Training Protocol

Data Augmentation. We first resize the image to the size of 512×512, and then we adopt data augmentation, including flip, rotation, Gaussian noise and Gaussian smoothing during training.

Table 3. Training protocols for our method.

Network initialization	Default normal initialization by pytorch
Batch size	4
Patch size	512×512
Total epochs	50
Optimizer	Adamw
Initial learning rate (lr)	6e-4
Lr decay schedule	Linear Learning Rate Decay
Training time	3.0 h
Loss function	$\mathcal{L}_{dice} + \mathcal{L}_{ce}$

5   Results and Discussion

5.1   Quantitative Results on Validation Set

To analyze the effect of the proposed method, we sample 40 labeled patches from the train set as a validation set and use the rest 360 patches to train our model. We report the mean **F1** (**mF1**) score as the primary metric, representing the average **F1** score of all cell classes. All the networks are built with pretrained SAM-ViT-H as the backbone and we evaluate the effect of CrossEntropy loss ($\mathcal{L}_{ce}$), Dropout, and Depthwise Separable Convolution, respectively. As shown in Table 4, each module we introduced contributes to improving the final results significantly. Through experimental observations, we found that incorporating the Cross-Entropy (CE) loss enhances the stability of the training process. Additionally, including Depthwise Separable Convolutions and Dropout improves the generalization capability of the model. To validate the efficacy of our SAM-ViT-H, we tested several different pre-trained backbones on the official validation set. Table 5 indicates that our implementation of SAM-ViT-H achieved the best performance. Furthermore, to validate the effectiveness of the LoRA fine-tuning approach employed in our study, we designed comparative experiments. As illustrated in Table 6, the LoRA fine-tuning method we utilized significantly outperforms other methods.

Table 4. Ablation study for our method. (DSC: depthwise separable convolution)

Method	Tissue miou	Cell miou	F1
Cell-Tissue-ViT	0.8404	0.5083	0.7370
Cell-Tissue-ViT w/o $\mathcal{L}_{ce}$	0.8378	0.5067	0.7327
Cell-Tissue-ViT w/ DSC	0.8443	0.5169	0.7442
Cell-Tissue-ViT w/ Dropout	**0.8480**	**0.5202**	**0.7480**

Table 5. Results on the Official *Validation* Set Under Different Pre-trained Backbone Protocols. (Imagenet means models pre-trained on ImageNet-21k dataset. B, L, H in the table refer to Base, Large, and Huge ViT versions)

Method	Imagenet $-$ VIT$_H$	SAM $-$ ViT$_B$	SAM $-$ ViT$_L$	SAM $-$ ViT$_H$
F1	0.7389	0.7441	0.7423	**0.7518**

Table 6. Results on the Official *Validation* Set Under Different Fine-Tuning Strategies. (Adapter refers to a fine-tuning approach using an MLP layer, placed after the multi-head attention, akin to [21])

Method	Adapter	LoRA
F1	0.7340	**0.7518**

5.2   Model Ensemble

We train four Tissue-CellViTs with the same architecture but different initializations respectively and merge them together in the prediction stage. The setting for the training protocol is presented in Table 3. To enhance the overall performance, we employ a dual-model ensemble approach. This approach involves averaging the outputs from the tissue branch of four individual models and then adding this average to the features extracted by the cell branch encoder. Subsequently, we similarly average the outputs from the cell branch of the four models to obtain the final predicted result. Table 7 and Table 8 respectively illustrate the performance of our method on the official validation and test sets. Additionally, Fig. 3 stands as a testament to the superiority of our approach, showcasing how the proposed Cell-Tissue-ViT achieves nuanced recognition of intricate cellular patterns within a vast field of view. The illustration captures how the Cell-only model falls short, often mistaking small, uniformly shaped cells for background elements due to its limited interpretive scope. In contrast, our method excels by integrating a comprehensive understanding of both cellular morphology and the surrounding tissue context. This dual consideration enables Cell-Tissue-ViT to discern and accurately classify cancerous areas that the Cell-only model overlooks.

Table 7. Results on the official *validation* set under different initialization protocols.

Method	C $-$ TViT$_1$	C $-$ TViT$_2$	C $-$ TViT$_3$	C $-$ TViT$_4$	Ensemble
F1	0.7307	0.7441	0.7511	0.7518	**0.7558**

Table 8. Results on the official *test* set under different initialization protocols.

Method	$C - TViT_1$	$C - TViT_2$	$C - TViT_3$	$C - TViT_4$	Ensemble
F1	0.7211	0.7208	0.7133	0.7185	**0.7243**

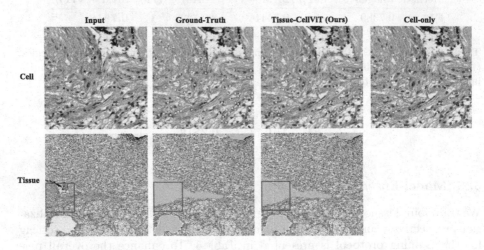

Fig. 3. Qualitative results between the proposed Cell-Tissue-ViT and the Standard-Cell-only model. Cell-Tissue-ViT demonstrates superior detection accuracy, benefiting from the integration of contextual tissue characteristics (yellow: tumor cells, blue: background cells, green: cancel area). (Color figure online)

6   Conclusion

In this paper, we propose an innovative ViT-based U-Net architecture, Tissue-CellViT, designed for enhanced cell detection in histopathological images. This approach adeptly integrates tissue contextuality and cellular detail through a sophisticated encoder-decoder mechanism, capitalizing on the strengths of the pre-trained SAM and LoRA optimization. Our method distinguishes itself by superior feature representation, achieving one of the highest **F1**-detection scores reported. Such a performance underscores our model's efficacy in interpreting complex biological structures, potentially revolutionizing computational pathology and fostering advancements in diagnostic precision.

References

1. Abels, E., et al.: Computational pathology definitions, best practices, and recommendations for regulatory guidance: a white paper from the digital pathology association. J. Pathol. **249**(3), 286–294 (2019)
2. Chollet, F.: Xception: deep learning with depthwise separable convolutions. In: Proceedings of the IEEE Conference on Computer Vision and Pattern Recognition, pp. 1251–1258 (2017)

3. Çiçek, Ö., Abdulkadir, A., Lienkamp, S.S., Brox, T., Ronneberger, O.: 3D U-net: learning dense volumetric segmentation from sparse annotation. In: Ourselin, S., Joskowicz, L., Sabuncu, M.R., Unal, G., Wells, W. (eds.) MICCAI 2016, Part II. LNCS, vol. 9901, pp. 424–432. Springer, Cham (2016). https://doi.org/10.1007/978-3-319-46723-8_49

4. Cutler, K.J., et al.: Omnipose: a high-precision morphology-independent solution for bacterial cell segmentation. Nat. Methods **19**(11), 1438–1448 (2022)

5. Hatamizadeh, A., et al.: Unetr: transformers for 3d medical image segmentation. In: Proceedings of the IEEE/CVF Winter Conference on Applications of Computer Vision, pp. 574–584 (2022)

6. Hörst, F., et al.: Cellvit: vision transformers for precise cell segmentation and classification. arXiv preprint arXiv:2306.15350 (2023)

7. Hu, E.J., et al.: Lora: low-rank adaptation of large language models. arXiv preprint arXiv:2106.09685 (2021)

8. Jia, M., et al.: Visual prompt tuning. In: Avidan, S., Brostow, G., Cissé, M., Farinella, G.M., Hassner, T. (eds.) ECCV 2022. LNCS, vol. 13693, pp. 709–727. Springer, Cham (2022). https://doi.org/10.1007/978-3-031-19827-4_41

9. Kirillov, A., et al.: Segment anything. arXiv preprint arXiv:2304.02643 (2023)

10. Mai, H., Sun, R., Zhang, T., Xiong, Z., Wu, F.: Dualrel: semi-supervised mitochondria segmentation from a prototype perspective. In: Proceedings of the IEEE/CVF Conference on Computer Vision and Pattern Recognition, pp. 19617–19626 (2023)

11. Pachitariu, M., Stringer, C.: Cellpose 2.0: how to train your own model. Nat. Methods **19**(12), 1634–1641 (2022)

12. Pan, Y., et al.: Adaptive template transformer for mitochondria segmentation in electron microscopy images. In: Proceedings of the IEEE/CVF International Conference on Computer Vision, pp. 21474–21484 (2023)

13. Ronneberger, O., Fischer, P., Brox, T.: U-net: convolutional networks for biomedical image segmentation. In: Navab, N., Hornegger, J., Wells, W.M., Frangi, A.F. (eds.) MICCAI 2015, Part III. LNCS, vol. 9351, pp. 234–241. Springer, Cham (2015). https://doi.org/10.1007/978-3-319-24574-4_28

14. Ryu, J., et al.: Ocelot: overlapped cell on tissue dataset for histopathology. In: Proceedings of the IEEE/CVF Conference on Computer Vision and Pattern Recognition, pp. 23902–23912 (2023)

15. Stringer, C., Wang, T., Michaelos, M., Pachitariu, M.: Cellpose: a generalist algorithm for cellular segmentation. Nat. Methods **18**(1), 100–106 (2021)

16. Sun, R., Li, Y., Zhang, T., Mao, Z., Wu, F., Zhang, Y.: Lesion-aware transformers for diabetic retinopathy grading. In: Proceedings of the IEEE/CVF Conference on Computer Vision and Pattern Recognition, pp. 10938–10947 (2021)

17. Sun, R., et al.: Appearance prompt vision transformer for connectome reconstruction. In: Elkind, E. (ed.) Proceedings of the Thirty-Second International Joint Conference on Artificial Intelligence, IJCAI-23, pp. 1423–1431. International Joint Conferences on Artificial Intelligence Organization (2023). https://doi.org/10.24963/ijcai.2023/158, main Track

18. Sun, R., Mai, H., Luo, N., Zhang, T., Xiong, Z., Wu, F.: Structure-decoupled adaptive part alignment network for domain adaptive mitochondria segmentation. In: Greenspan, H., et al. (eds.) MICCAI 2023. LNCS, vol. 14223, pp. 523–533. Springer, Cham (2023). https://doi.org/10.1007/978-3-031-43901-8_50

19. Swiderska-Chadaj, Z., et al.: Learning to detect lymphocytes in immunohistochemistry with deep learning. Med. Image Anal. **58**, 101547 (2019)

20. Wangkai, L., et al.: MauNet: modality-aware anti-ambiguity u-net for multi-modality cell segmentation. In: Competitions in Neural Information Processing Systems, pp. 1–12. PMLR (2023)
21. Wu, J., et al.: Medical SAM adapter: adapting segment anything model for medical image segmentation. arXiv preprint arXiv:2304.12620 (2023)
22. Zhang, K., Liu, D.: Customized segment anything model for medical image segmentation. arXiv preprint arXiv:2304.13785 (2023)
23. Zhang, Y., Jiao, R.: How segment anything model (sam) boost medical image segmentation? arXiv preprint arXiv:2305.03678 (2023)
24. Zhou, Z., Siddiquee, M.M.R., Tajbakhsh, N., Liang, J.: Unet++: redesigning skip connections to exploit multiscale features in image segmentation. IEEE Trans. Med. Imaging **39**(6), 1856–1867 (2019)

Generating BlobCell Label from Weak Annotations for Precise Cell Segmentation

Suk Min Ha[1], Young Sin Ko[1,2(✉)], and Youngjin Park[1(✉)]

[1] AI Research Institute, Seegene Medical Foundation, Seoul, Korea
{hasukmin12,noteasy,youngjpark}@mf.seegene.com
[2] Pathology Center, Seegene Medical Foundation, Seoul, Korea

Abstract. Cell segmentation in histopathological image analysis is critical for identifying cancer cells and predicting disease severity. However, manual cell labeling is time-consuming. Many experiments speed up the generation of cell data by annotating central cell points and classes, generating cell segmentation labels with a fixed radius. However, the accuracy of this method depends on the specified given radius, which is problematic due to the variety of cell sizes and shapes, including elongated ovals and linear shapes. The use of a fixed radius is considered in-accurate and unreasonable. To address this, we propose BlobCell labeling, which uses blob extraction for accurate cell labeling based on central coordinates, resulting in a **+9.02%** improvement in the dice score. Furthermore, to improve cell detection from cell segmentation results such as the proposed challenge baseline [1], we designed a new network architecture that utilizes Blob-Cell information within the Injection model structure, we achieved a significant performance improvement of **+12.11%** in mF1 score on the test set.

Keywords: Cell Detection · Weak annotation · Cell-Tissue Interaction · Blob detection algorithm · Deep learning segmentation · Weak annotation · Pathology

1 Introduction

Cell detection is a critical step in histopathological image analysis. It is used to identify cancer cells and other abnormalities, which can help predict disease severity and patient prognosis [7–9]. Deep learning has emerged as a promising approach to cell detection [7–9] and is likely to play an increasingly important role in histopathological image analysis. However, previously proposed methods may rely on appearance details without understanding the broader context [10]. The broader context can provide a deeper understanding of the spatial relationships between cells and high-level tissue structures, that can lead to more accurate cell detection.

The OCELOT Grand Challenge [1], "OCELOT 2023: Cell Detection from Cell-Tissue Interaction", was launched as part of MICCAI 2023 to address this pressing issue. This dataset includes annotations for both cells and tissues in patches of different field of view (FoV), including both large and small patches with overlapping regions. The tissue

---

The original version of the chapter has been revised. Some errors occurred. This was corrected. A correction to this chapter can be found at
https://doi.org/10.1007/978-3-031-55088-1_16

S.-A. Ahmadi and S. Pereira (Eds.): MICCAI 2023, LNCS 14373, pp. 161–170, 2024.
https://doi.org/10.1007/978-3-031-55088-1_15

dataset contains annotations for cancer areas, which are marked by segmentation. On the other hand, the cell dataset contains background cells and tumor cells, each identified by their central coordinates and classes, all stored as annotations in JSON files. To facilitate joint training of these two datasets and establish a relationship between them, the central coordinates of the cells as annotated in the JSON files are used to generate cell segmentation labels.

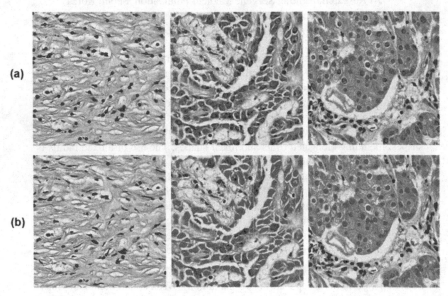

Fig. 1. (a) Defines cell radius as 2.33 μm; (b) Newly proposed BlobCell label generation method; Blue represents BC (Background Cell), while red signifies TC (Tumor Cell). (Color figure online)

Manually labeling individual cells is a time-consuming and costly work. In many experiments, cell data is quickly generated by labeling only the central points of cells and specifying their classes [11, 12]. A cell segmentation label is then created by generating a circle of a given radius around these central points and defining it as the cell segmentation label [2]. However, the performance of this approach is highly dependent on the given radius value. Cells have a wide range of sizes and shapes, from elongated ovals to shapes that resemble straight lines rather than simple circles. Consequently, generating cell labels with a fixed radius may not be considered accurate and may be seen as an unreasonable approach.

Recognizing this challenge, we consider the creation of accurate cell labels to be one of the most important aspects of this competition. As a solution, we propose the BlobCell label, a more rational approach to cell labeling. This method doesn't just rely on generating labels based on a specific radius, but instead uses Gaussian blur to extract blobs from the image. We define the blob containing the given cell center location (x, y) as the cell segmentation label. Figure 1 illustrates the form of our proposed BlobCell approach compared to the approach that uses a fixed cell radius to generate cell segmentation labels.

This approach allows us to generate more accurate cell labels, even for cells with elliptical or linear shapes. Using this method, we have demonstrated that it is possible to generate accurate cell segmentation labels using only cell center location, resulting in a remarkable $+9.02\%$ improvement in segmentation performance.

The proposed competition places considerable emphasis on categorizing cells as BC or TC, and to address this they propose a method that incorporates tissue information. However, when performing the actual inference, the performance of the segmentation of the cells themselves appears to be suboptimal. To address this, we have developed a structure that includes an additional branch for precise cell segmentation using BlobCell labels. This addition assists the main model in locating cells more accurately. Overall structure shown in Fig. 4, and this design has resulted in an improvement of both $+19.56\%$ in segmentation and $+12.11\%$ in detection performance on the test set (Table 2). Code is available at the following link:

https://github.com/hasukmin12/OCELOT_2023_BlobCell_Method.git.

2 Related Work

2.1 Generate Cell Segmentation Labels from Cell Center Locations

Creating segmentation labels by manual drawing the cell is time consuming and costly. Marking only the central points of cells and providing their classes significantly reduces these resources. For each annotated cell center location (x, y), it was observed that creating a cell segmentation label with a specific radius, using U-Net [14] for training, and finding the cell centroid shows superior performance compared to learning by cell detection methods such as YOLO [11, 12] and LSM [2]. In the proposed method [2], regions approximating the cell body and membrane of each lymphocyte were defined with radii of 2.4 μm and 2.88 μm, respectively.

The OCELOT cell dataset consists of micron-per-pixel values ranging from 0.1942 to 0.1944. In our experiments, we provided radius values close to the proposed range (2.33 μm to 2.91 μm) and compared the performance with our proposed BlobCell method. As the defined radius value approaches 2.33 μm, it primarily represents the nucleus, whereas as it approaches 2.91 μm, it represents the entire cell.

2.2 Blob Extraction

The use of blobs for cell extraction is one of the methods used to extract cells from pathology images [6]. A "Blob" refers to a contiguous and distinguishable region within an image. Cells in pathology images typically have distinct colors and shapes that can be identified as blobs by blob detection algorithms.

2.3 Tissue Injection Model

Pathologists do not make diagnoses at a single magnification. They often zoom in and out to understand tissue-level structures. Therefore, it appears that, similar to human pathologists, AI can improve diagnostic accuracy by making the most of tissue-level structures rather than performing cell detection at a single magnification level.

In the challenge paper [1], an existing model was developed to segment tumor areas within the tissue at a lower level of magnification to allow for a broader context. The tissue-level segmentation results were then used to crop and upsampling as the corresponding FoV of the cell patches, followed by channel-wise concatenation with the cell patch. By incorporating tumor area information into the cell patches with Pred-to-inter-2 approach, a significant improvement in mean F1 score performance was observed as + **7.69%** and **+9.76%** [1] in the validation set and test set respectively on the OCELOT dataset.

3 Method

3.1 BlobCell Label

Cell nuclei in pathology images have distinctive features that can be extracted using blob detection algorithms. The process of cell extraction using blobs can be divided into the following steps.

1) **Convert the image to grayscale**: to remove color information, making it easier to identify the shapes of the cells.
2) **Apply Gaussian blur and use thresholding**: Gaussian blur is used to remove noise on image and thresholding is used to convert the image to a binary format, making it easier to identify blobs. Functions are used from cv2 (OpenCV).
3) **Use blob detection algorithms to extract blobs**: Utilizing the findContours function from cv2 (OpenCV) to implement the blob detection algorithm and extract blobs from a binarized image.
4) **Only blobs containing cell center location (x, y) from the extracted blobs are retained**: Blobs that do not contain cell center location are removed. For cells whose cell center location is not inside the blob, a circle with a radius of 2.33 μm is drawn to generate the cell segmentation label.

The BlobCell generation process can be observed in Fig. 2. And Fig. 3 shows the results of the BlobCell label compared to the approach that uses a fixed cell radius to generate cell segmentation labels.

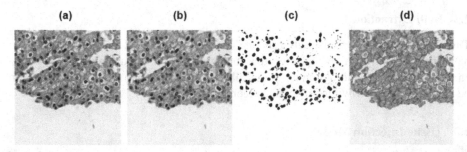

Fig. 2. BlobCell generation process. (a) original image; (b) grayscale applied; (c) Gaussian Blur and Threshold applied; (d) contours are extracted and identification of the blobs.

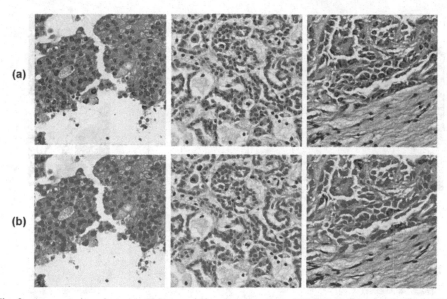

Fig. 3. A comparison between the conventional approach and BlobCell. **Blue: BC; Red: TC; Green: Blobs that are not included in the cell center location;** (a) The conventional approach, generating labels by drawing a circle based on a specific given radius; (b) The BlobCell labels, the green color indicates the blobs that are not included in the cell center location, and will be eliminated during the process. (Color figure online)

3.2 BlobCell-Tissue Prediction Injection Models

Using the previously introduced BlobCell, we present an optimized segmentation model designed for cell segmentation. We prioritized providing more spatial information about cells. The competition considers crucial focus in the accurate classification of cells as BC or TC. However, in the actual inference it seems that there are more errors in finding the cells themselves. This may be due to problems with different cell shapes, or domain-specific issues related to different organs. With a primary focus on addressing this issue, we designed a new structure that includes an additional branch dedicated to providing accurate cell location information, as shown in Fig. 4. This extension helps the main model to detect cells more accurately.

The proposed structure is composed of three models, BlobCell Model, Tissue Model, Main Model, and all models consist of SE ResNext [4, 5] encoder and DeepLabv3 + decoder [13]. All the encoders used the pretrained weights from ImageNet.

In the first branch, the cell patch (small FoV) images are fed into the BlobCell model to produce a BlobCell segmentation output. This model presents all BlobCells into a single class without defining whether they are BC or TC. It is used to provide accurate cell location information to the main model based on the refined BlobCell segmentation results.

The second branch, called the Tissue Model, is responsible for segmenting the tumor area using tissue information from large Field of View patches, following the approach outlined in the competition paper [1]. The segmentation output is then cropped and

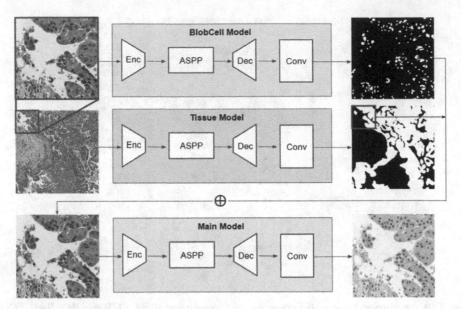

Fig. 4. Structure of the proposed model. Based on the results predicted by the BlobCell Model and the Tissue Model, a 5-channel image is created and the Main Model trains and evaluates it.

upsampled to match the small FoV area. Finally, the BlobCell segmentation output from the first branch and the cropped tissue segmentation result from the second branch are channel-wise concatenated with the original cell patch (small FoV) image. This produces a 5-channel image that trains the main model to perform cell segmentation with a specified radius of 2.33 μm. Figure 5 shows the results of each model during the inference process.

The reason for configuring the main model to output cells with a radius of 2.33 μm, rather than deriving the BlobCell output directly, is that while BlobCell provides accurate cell segmentation, extracting cell center locations from BlobCell leads to inaccurate results. (Table 1). Therefore, to be consistent with the ultimate goal of the competition, which is cell detection, and to achieve a significant increase in the mean F1 score, the final results are designed to produce a cell output with a specified radius.

The aim of this approach was to improve the accuracy of cell location detection. The results confirmed our expectations, showing an improvement of **+15.96%** and **+15.94%** in the mF1 score on the validation and test datasets, respectively. (Table 2).

3.3 Implementation Details

The network was trained for 100 epochs with a patch size of (1024, 1024). Dice cross-entropy loss [16] was used as the loss function and AdamW [15] as the optimizer. The training was performed with a batch size of 8, using an NVIDIA A100 80GB GPU. The data proposed in the challenge come from different organs. Therefore, it was important to build a model with multi-organ compatibility. Therefore, we set 20% of each organ as a validation set to allow all organs to be trained and evaluated at least once, and applied Stain Normalization [3] to increase multi-organ compatibility.

4   Experiment and Results

Compared the performance of Cell-only structure using the closest proposed radius values (2.4 μm and 2.88 μm) [2], BlobCell method and OCELOT's result. The Dice score, also known as the Dice coefficient, is a statistic used to measure the similarity or agreement between two sets. It is commonly used in the context of evaluating the performance of image segmentation algorithms. The mF1 score, as the primary metric used for evaluation in the challenge, is a metric used to assess the overall performance of a classification model, particularly in multi-class classification tasks. It measures the balance between precision and recall for each class in the classification problem and calculates the mean of these individual class F1 scores to provide an aggregate measure of the model's performance. The mF1 score serves as the primary metric for this research and is the most important metric in the challenge.

Performance varied considerably as the defined radius values changed, as measured by the dice score. This change can be attributed to imperfect and inconsistent label generation. Overall, larger defined cell radii resulted in higher Dice scores but lower mF1 scores. The mF1 score was highest when a 2.33 μm radius was used, outperforming the OCELOT experimental results [1] by +0.2%, +1.59% in mF1 score for validation and test sets. As a result, we standardized the specific cell radius to 2.33 μm.

The use of our proposed BlobCell labeling method led to a significant improvement in segmentation performance. Our approach resulted in a remarkable improvement of approximately +9.02%, +8.68% on the validation and test set in terms of dice score than 2.33 μm cell, as detailed in Table 1. However, it's important to note that in the OCELOT challenge, participants were evaluated primarily on their ability to accurately locate and measure cell centroids using the mean F1 score, rather than focusing on cell segmentation performance. In this context, the BlobCell method experienced a significant drop in performance, with reductions of approximately −11.62%, −10.97% on the validation and test set.

The observed problem seems to be due to the incompatibility of the BlobCell method with OCELOT's "find_cells" algorithm for extracting cell center positions from segmentation output. In addition, the use of BlobCell often results in enlarged cell allocation outputs, making the measurement of cell center positions potentially disturbing. As a result, while the BlobCell method helps to achieve accurate cell segmentation, it falls short when it comes to accurately measuring cell center locations. So we have shifted our research focus to developing a BlobCell injection model, rather than using BlobCell as a final output [2],

The paper presented in the challenge compared the cell-only approach with the Pred-to-input method [1]. Significant improvements were observed in the evaluation results on the validation and test sets, with mF1 scores increasing by +4.49% and +5.21% respectively (Table 2).

When experimenting with a self-designed Tissue injection model based on the the proposed paper [1], the mF1 scores showed improvements of +3.72% and +3.83%. The injected tissue information was provided as Ground Truth (GT), similar to the approach in the referenced paper [1], and the results were in close agreement with the paper.

Training the proposed BlobCell-Tissue injection model resulted in a significant increase in Dice scores of +20.33% and +19.56%, and mF1 scores also showed a significant improvement of +12.24% and +12.11% compared to the Tissue injection model. This highlights the difficulty in cell detection tasks is not classification, but rather cell localization.

When the performance of the injected model was suboptimal, as this could have a significant impact on the final result. To account this, we conducted experiments where the information from BlobCell was injected as a GT during inference.

But, due to time constraints, we implemented the Pred-to-input approach by concatenating all input images into a main model input. We believe that after further refinement of the BlobCell and Tissue models, implementation of the Pred-to-inter-2 approach could lead to better results.

Table 1. Cell Segmentation Dice score, and Cell detection mean F1 scores per various radius setting, in Cell-only structure. Results trained with the specified cell radius label; Results using BlobCell; Results of the OCELOT team;

Cell radius		OCELOT	2.33 μm	2.52 μm	2.72 μm	2.91 μm	BlobCell
Val set	Dice score	–	57.35	59.42	62.25	61.38	**71.27**
	mF1-score	68.87	**69.07**	68.11	68.72	68.53	57.45
Test set	Dice score	–	57.30	59.53	61.76	62.47	**70.44**
	mF1-score	64.44	**66.03**	64.85	63.34	64.92	55.06

Table 2. Cell Segmentation Dice score, and Cell detection mean F1-scores per model. Cell-only model with best performing radius value (2.33 μm); Tissue injection model (Pred-to-input); Proposed BlobCell-Tissue injection model (Pred-to-input); OCELOT team's Cell-only model; OCELOT's Tissue Injection model (Pred-to-input)

Method		OCELOT Cell-only	OCELOT Tissue Inject	Cell-only	Tissue injection	BlobCell-Tissue injection
Val set	Dice score	–	–	57.35	59.72	**80.05**
	mF1-score	68.87	73.36	69.07	72.79	**85.03**
Test set	Dice score	–	–	57.30	59.45	**79.01**
	mF1-score	64.44	69.65	66.03	69.86	**81.97**

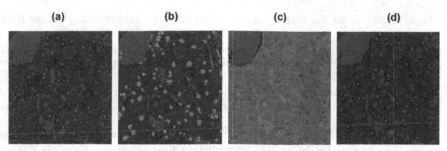

Fig. 5. The test set inference process of the BlobCell-Tissue Injection model. **Blue: BC; Red: TC; Green: BlobCell prediction;** (a) GT; (b) BlobCell Model Prediction; (c) Tissue Model Prediction; (d) Final result of BlobCell-Tissue Injection model (Color figure online)

5 Conclusion

Our approach has shown that it's possible to generate accurate cell segmentation labels using only cell center coordinates. Furthermore, these labels result in more accurate cell segmentation compared to existing methods. Our approach has shown that it's possible to generate accurate cell segmentation labels using only cell center coordinates. Furthermore, these labels result in more accurate cell segmentation compared to existing methods. While BlobCell significantly improves cell segmentation, it may appear to compromise cell detection performance due to the challenges of accurately determining cell center coordinates.

However, our BlobCell-Tissue injection model was specifically designed to outperform the mF1 score of the Tissue injection model proposed in the challenge paper [1] by providing more accurate cell locate information and classification. And this result emphasises that the primary difficulty in cell detection is localization rather than classification.

Despite the promising experimental results, all experiments were performed with provided ground truth (GT) for tissue and BlobCell. We did not develop fully optimised models for tissue segmentation and BlobCell, which resulted in performance degradation and fell short of the high level of performance suggested in Table 2 of the challenge. In addition, due to time constraints, we implemented the pred-to-input approach by concatenating all input images into the main model input. We anticipate that further refinement of the BlobCell and Tissue models, together with implementation of the Pred-to-inter-2 approach, could lead to improved results.

References

1. Ryu, J., et al.: OCELOT: overlapped cell on tissue dataset for histopathology. In: Proceedings of the IEEE/CVF Conference on Computer Vision and Pattern Recognition, pp. 23902–23912 (2023)
2. Swiderska-Chadaj, Z., et al.: Learning to detect lymphocytes in immunohistochemistry with deep learning. Med. Image Anal. **58**, 101547 (2019)
3. Michielli, N., et al.: Stain normalization in digital pathology: clinical multi-center evaluation of image quality. J. Pathol. Inform. **13**, 100145 (2022)

4. Hu, J., Shen, L., Sun, G.: Squeeze-and-excitation networks. In: Proceedings of the IEEE Conference on Computer Vision and Pattern Recognition, pp. 7132–7141 (2018)
5. Xie, S., Girshick, R., Dollár, P., Tu, Z., He, K.: Aggregated residual transformations for deep neural networks. In: Proceedings of the IEEE Conference on Computer Vision and Pattern Recognition, pp. 1492–1500 (2017)
6. Al-Kofahi, Y., Zaltsman, A., Graves, R., Marshall, W., Rusu, M.: A deep learning-based algorithm for 2-D cell segmentation in microscopy images. BMC Bioinform. **19**(1), 1–11 (2018)
7. Lal, S., Das, D., Alabhya, K., Kanfade, A., Kumar, A., Kini, J.: NucleiSegNet: robust deep learning architecture for the nuclei segmentation of liver cancer histopathology images. Comput. Biol. Med.. Biol. Med. **128**, 104075 (2021)
8. Qu, H., et al.: Weakly supervised deep nuclei segmentation using partial points annotation in histopathology images. IEEE Trans. Med. Imaging **39**(11), 3655–3666 (2020)
9. Zhao, B., et al.: Triple U-net: hematoxylin-aware nuclei segmentation with progressive dense feature aggregation. Med. Image Anal. **65**, 101786 (2020)
10. Tokunaga, H., Teramoto, Y., Yoshizawa, A., Bise, R.: Adaptive weighting multi-field-of-view CNN for semantic segmentation in pathology. In: Proceedings of the IEEE/CVF Conference on Computer Vision and Pattern Recognition, pp. 12597–12606 (2019)
11. Redmon, J., Farhadi, A.: Yolov3: an incremental improvement. arXiv preprint arXiv:1804.02767 (2018)
12. Nair, L.S., Prabhu, R., Sugathan, G., Gireesh, K.V., Nair, A.S.: Mitotic nuclei detection in breast histopathology images using Yolov4. In: 2021 12th International Conference on Computing Communication and Networking Technologies (ICCCNT), pp. 1–5. IEEE, July 2021
13. Chen, L.-C., Zhu, Y., Papandreou, G., Schroff, F., Adam, H.: Encoder-decoder with atrous separable convolution for semantic image segmentation. In: Ferrari, V., Hebert, M., Sminchisescu, C., Weiss, Y. (eds.) ECCV 2018. LNCS, vol. 11211, pp. 833–851. Springer, Cham (2018). https://doi.org/10.1007/978-3-030-01234-2_49
14. Ronneberger, O., Fischer, P., Brox, T.: U-net: convolutional networks for biomedical image segmentation. In: Navab, N., Hornegger, J., Wells, W.M., Frangi, A.F. (eds.) MICCAI 2015. LNCS, vol. 9351, pp. 234–241. Springer, Cham (2015). https://doi.org/10.1007/978-3-319-24574-4_28
15. Loshchilov, I., Hutter, F.: Decoupled weight decay regularization. arXiv preprint arXiv:1711.05101 (2017)
16. Milletari, F., Navab, N., Ahmadi, S.A.: V-net: fully convolutional neural networks for volumetric medical image segmentation. In: 2016 Fourth International Conference on 3D Vision (3DV), pp. 565–571. IEEE, October 2016

Correction to: Generating BlobCell Label from Weak Annotations for Precise Cell Segmentation

Suk Min Ha, Young Sin Ko, and Youngjin Park

Correction to:
Chapter 16 in: S.-A. Ahmadi and S. Pereira (Eds.): *Graphs in Biomedical Image Analysis, and Overlapped Cell on Tissue Dataset for Histopathology,* **LNCS 14373,**
https://doi.org/10.1007/978-3-031-55088-1_15

In the original version of this paper some errors occurred. This was corrected.

The updated version of this chapter can be found at
https://doi.org/10.1007/978-3-031-55088-1_15

Author Index

S.-A. Ahmadi and S. Pereira (Eds.): MICCAI 2023, LNCS 14373, pp. 171–172, 2024.
https://doi.org/10.1007/978-3-031-55088-1

Printed in the United States
by Baker & Taylor Publisher Services